Nonlinear Systems

Stability, Dynamics and Control

Nonlinear Systems

Stability, Dynamics and Control

Guanrong Chen

City University of Hong Kong, China

W♭ World Scientific

NEW JERSEY · LONDON · SINGAPORE · BEIJING · SHANGHAI · HONG KONG · TAIPEI · CHENNAI · TOKYO

Published by

World Scientific Publishing Co. Pte. Ltd.

5 Toh Tuck Link, Singapore 596224

USA office: 27 Warren Street, Suite 401-402, Hackensack, NJ 07601

UK office: 57 Shelton Street, Covent Garden, London WC2H 9HE

Library of Congress Control Number: 2023028635

British Library Cataloguing-in-Publication Data
A catalogue record for this book is available from the British Library.

NONLINEAR SYSTEMS
Stability, Dynamics and Control

ISBN 978-981-127-898-3 (hardcover)
ISBN 978-981-127-899-0 (ebook for institutions)
ISBN 978-981-127-900-3 (ebook for individuals)

For any available supplementary material, please visit
https://www.worldscientific.com/worldscibooks/10.1142/13488#t=suppl

Dedicated to the memory of
Professor Gennady Alexeyevich Leonov (1947–2018)

Preface

Nonlinear systems constitute a fundamental subject for study in systems engineering. The subject has been extensively investigated by both nonlinear control and nonlinear dynamics communities, whereas the focus is usually very different, on controllers design and dynamics analysis, respectively. The last two decades have witnessed gradual merging of control theory and dynamics analysis, but not to the extent of controlling nonlinear dynamics such as bifurcations and chaos. This monograph is an attempt to fill the gap to a certain extent while presenting a rather comprehensive coverage of the nonlinear systems theory in a self-contained and hopefully easily-readable manner.

This introductory treatise is not intended to be a research reference with the state-of-the-art theories and techniques presented, nor as a very comprehensive handbook, given that there are already many available in the market today. It is written for self-study and, in particular, as an elementary textbook that can be taught in a one-semester course at the advanced undergraduate level or entrance level of graduate curricula focusing on nonlinear systems — both control theory and dynamics analysis.

The main contents of the book comprise systems stability (Chapters 2–4), bifurcation and chaos dynamics (Chapter 5) and controllers design (Chapter 6), for both continuous-time and discrete-time settings. In particular, it discusses the special topics on bifurcation control and chaos control at the end of the last chapter.

This monograph is presented in a textbook style, in which most contents are elementary with some classical results and popular examples taken or modified from the existing literature, which might have also appeared in some other introductory textbooks. Since this is not a survey, a long list of related references is not included, yet appreciation to the various original sources are indicated. Throughout the book, to keep its contents at an elementary level, some advanced theories are presented without detailed proofs, merely for the completeness of the relevant discussions. This kind of contents and exercises are marked by * for indication.

To this end, I would like to especially thank Dr. Yi Jiang from the City University of Hong Kong for helping proof-read the entire manuscript, and thank Ms. Lakshmi Narayanan from the World Scientific Publishing Company for her support and assistance.

Guanrong Chen
Hong Kong, 2023

Contents

Notation

R^n	Space of n-dimensional real vectors
C^1	Space of continuously differential functions
$A^\top$	Transpose of matrix A
$\lVert \cdot \rVert$	Euclidean norm (L_2-norm)
$\circ$	Composite operation of two maps
$\mathrm{Re}\{\lambda\}$	Real part of eigenvalue λ
$\lambda_{\max}(A)$	Maximum eigenvalue of matrix A
$\lambda_{\min}(A)$	Minimum eigenvalue of matrix A
$\mathrm{diag}\{\cdot\}$	Diagonal matrix
$\mathrm{trace}(A)$	Trace of matrix A
$\exp\{x\}$	Exponential function e^x
$\inf\{\cdot\}$	Infimum

Chapter 1

Nonlinear Systems: Preliminaries

A *nonlinear system* in the mathematical sense refers to a set of nonlinear equations, which can be algebraic, difference, differential, integral, functional, and operator equations, or a combination of some of them. A nonlinear system is used to describe a physical device or process that otherwise cannot be well defined by a set of linear equations of any kind, although a linear system is considered as a special case of a nonlinear system. *Dynamical system*, on the other hand, is used as a synonym of a mathematical or physical system, where the output behavior evolves with time and sometimes with other varying system parameters as well. The system responses, or behaviors, of a dynamical system is referred to as *system dynamics*.

1.1 A Typical Nonlinear Dynamical Model

A representative mathematical model of nonlinear dynamical systems is the pendulum equation. The study of pendula can be traced back to as early as Christian Huygens who investigated in 1665 the perfect synchrony of two identical pendulum clocks that he invented in 1656. He then reported his findings to the Royal Society of The Netherlands [Huygens (1665)].

A simple and idealized pendulum consists of a volumeless ball and a rigid and massless rod, which is connected to a pivot, as shown in Fig. 1.1. In this figure, ℓ is the length of the rod, m is the mass of the ball, g is the constant of gravity acceleration, $\theta = \theta(t)$ is the angle of the rod with respect to the vertical axis, and $f = f(t)$ is the resistive force applied to the ball. The straight down position finds the ball at rest; but if it is displaced by an angle from the reference axis and then let go, it will swing back and forth on a circular arc within a vertical plane to which it is confined.

For general purpose of mathematical analysis of the pendulum, a basic

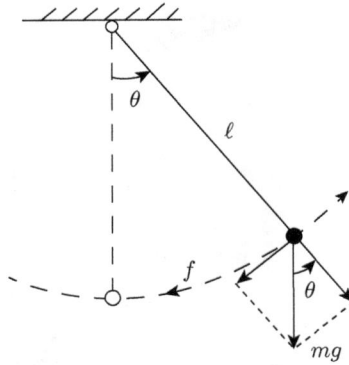

Fig. 1.1   Schematic diagram of a pendulum model.

assumption is that the resistive force f is proportional to the velocity along the arc of the motion trajectory of the volumeless ball, i.e. $f = \kappa \dot{s}$, where $\kappa \geq 0$ is a constant and $s = s(t) = \ell\,\theta(t)$ is the ball-traveled arc length measured from the vertical reference axis.

It follows from Newton's second law of motion that

$$m\,\ddot{s} = -mg\,\sin(\theta) - \kappa\,\dot{s}\,,$$

or

$$\ddot{\theta} + \frac{\kappa}{m}\,\dot{\theta} + \frac{g}{\ell}\,\sin(\theta) = 0\,. \tag{1.1}$$

This is the idealized and damped pendulum equation, which is nonlinear due to the involvement of the sine function. When $\kappa = 0$, i.e. $f = 0$, it becomes the undamped pendulum equation

$$\ddot{\theta} + \frac{g}{\ell}\,\sin(\theta) = 0\,. \tag{1.2}$$

To this end, by introducing two new variables,

$$x_1 = \theta \qquad \text{and} \qquad x_2 = \dot{\theta}\,,$$

the damped pendulum equation can be rewritten in the following state-space form:

$$\dot{x}_1 = x_2\,,$$

$$\dot{x}_2 = -\frac{\kappa}{m}\,x_2 - \frac{g}{\ell}\,\sin(x_1)\,. \tag{1.3}$$

Here, the variables $x_1 = x_1(t)$ and $x_2 = x_2(t)$ are called the *system states*, for they describe the physical states, namely the angular position and angular velocity respectively, of the pendulum.

From elementary physics, it is known that the pendulum state vector $\mathbf{x}(t) = [x_1(t) \ x_2(t)]^\top$ is periodic. In general, even in a higher-dimensional case, a state $\mathbf{x}(t)$ of a dynamical system is a *periodic solution* if it is a solution of the system and moreover satisfies $\mathbf{x}(t + t_p) = \mathbf{x}(t)$ for some constant $t_p > 0$. The least value of such t_p is called the (*fundamental*) *period* of the periodic solution, while the solution is said to be t_p-*periodic*.

Although conceptually straightforward and formally simple, this pendulum model has many important and interesting properties. This representative model of nonlinear systems will be frequently referred to, not only within this chapter but also throughout the book.

1.2 Autonomous Systems and Map Iterations

The pendulum model (1.3) is called an *autonomous system*, in which there is no independent (or separated) time variable t, other than a time variable as the system states, anywhere in the model formulation. On the contrary, the following forced pendulum is *nonautonomous*:

$$\dot{x}_1 = x_2 \,,$$
$$\dot{x}_2 = -\frac{\kappa}{m}\, x_2 - \frac{g}{\ell}\, \sin(x_1) + h(t) \,, \tag{1.4}$$

since t exists as a variable in the function $h(t)$, independent of the system states, which in this example is an external force applied to the pendulum. One example of such a force input is $h(t) = a\,\cos(\omega t)$, which will change the angular acceleration of the pendulum, where a and ω are some constants.

The forced pendulum (1.4) has a time variable, t, within the external force term $h(t)$, which may not be shown as the time variables in the system states x_1 and x_2 for brevity. However, if $h(t) = a\,\cos(\theta(t))$, then the forced pendulum is considered to be autonomous because the time variable in the force term becomes the time variable of the system state $x_1(t)$. In the latter case, the external input should be denoted as $h(x_1)$ instead of $h(t)$ in the system equations.

In general, an n-dimensional autonomous system is described by

$$\dot{\mathbf{x}} = \mathbf{f}(\mathbf{x}; \mathbf{p}) \,, \qquad \mathbf{x}_0 \in R^n \,, \qquad t \in [t_0, \infty) \,, \tag{1.5}$$

and a nonautonomous system is expressed as

$$\dot{\mathbf{x}} = \mathbf{f}(\mathbf{x}, t; \mathbf{p}) \,, \qquad \mathbf{x}_0 \in R^n \,, \qquad t \in [t_0, \infty) \,, \tag{1.6}$$

where $\mathbf{x} = \mathbf{x}(t) = [x_1(t) \ \cdots \ x_n(t)]^\top$ is the *state vector*, $\mathbf{x}_0$ is the *initial state* with *initial time* $t_0 \geq 0$, $\mathbf{p}$ is a vector of *system parameters*, which can be varied but are independent of time, and

$$\mathbf{f} = \begin{bmatrix} f_1 \\ \vdots \\ f_n \end{bmatrix} = \begin{bmatrix} f_1(x_1, \ldots, x_n) \\ \vdots \\ f_n(x_1, \ldots, x_n) \end{bmatrix}$$

is called the *system function* or *vector field*.

In a well-formulated mathematical model, the system function should satisfy some defining conditions such that the model, for example (1.5) or (1.6), has a unique solution for each initial state $\mathbf{x}_0$ in a region of interest, $\Omega \subseteq R^n$, and for each permissible set of parameters $\mathbf{p}$. According to the elementary theory of ordinary differential equations, this is ensured if the function $\mathbf{f}$ satisfies the *Lipschitz condition*

$$||\mathbf{f}(\mathbf{x}) - \mathbf{f}(\mathbf{y})|| \leq \alpha \, ||\mathbf{x} - \mathbf{y}||$$

for all $\mathbf{x}$ and $\mathbf{y}$ in Ω that satisfy the system equation, and for some constant $\alpha > 0$, called the *Lipschitz constant*. Here and throughout the book, $|| \cdot ||$ denotes the standard Euclidean norm (the "length") of a vector, i.e. the L_2-norm. This general setting, i.e. with the fulfillment of some necessary defining conditions for a given mathematical model, will not be repeatedly mentioned and described later on, for simplicity of presentation.

Sometimes, an n-dimensional continuous-time dynamical system is described by a time-varying map,

$$F_c(t): \quad \mathbf{x} \to \mathbf{g}(\mathbf{x}, t; \mathbf{p}), \qquad \mathbf{x}_0 \in R^n, \qquad t \in [t_0, \infty), \qquad (1.7)$$

or, by a time-invariant map,

$$F_c: \quad \mathbf{x} \to \mathbf{g}(\mathbf{x}; \mathbf{p}), \qquad \mathbf{x}_0 \in R^n, \qquad t \in [t_0, \infty). \qquad (1.8)$$

These two maps, in the continuous-time case, take a function to a function; so by nature they are *operators*, which however will not be further studied in this book.

For the discrete-time case, with similar notation, a nonlinear dynamical system is either described by a time-varying difference equation,

$$\mathbf{x}_{k+1} = \mathbf{f}_k(\mathbf{x}_k; \mathbf{p}), \qquad \mathbf{x}_0 \in R^n, \qquad k = 0, 1, \ldots, \qquad (1.9)$$

where the subscript of $\mathbf{f}_k$ signifies the dependence of $\mathbf{f}$ on the discrete time variable k, or described by a time-invariant difference equation,

$$\mathbf{x}_{k+1} = \mathbf{f}(\mathbf{x}_k; \mathbf{p}), \qquad \mathbf{x}_0 \in R^n, \qquad k = 0, 1, \ldots. \qquad (1.10)$$

Also, it may be described either by a time-varying map,

$$F_d(k): \quad \mathbf{x}_k \to \mathbf{g}_k(\mathbf{x}_k; \mathbf{p}), \qquad \mathbf{x}_0 \in R^n, \qquad k = 0, 1, \ldots, \qquad (1.11)$$

or by a time-invariant map,

$$F_d: \quad \mathbf{x}_k \rightarrow \mathbf{g}(\mathbf{x}_k; \mathbf{p}), \qquad \mathbf{x}_0 \in R^n, \qquad k = 0, 1, \ldots . \qquad (1.12)$$

These discrete-time maps, particularly the time-invariant ones, are very important and convenient for the study of system dynamics; they will be further discussed in the book later on.

For a system given by a time-invariant difference equation, repeatedly iterating the system function $\mathbf{f}$ leads to

$$\mathbf{x}_k = \underbrace{\mathbf{f} \circ \cdots \circ \mathbf{f}}_{k \text{ times}} (\mathbf{x}_0) := \mathbf{f}^k(\mathbf{x}_0), \qquad (1.13)$$

where "$\circ$" denotes the composition operation of two functions or maps. Similarly, for a map F_d, repeatedly iterating it backwards yields

$$\mathbf{x}_k = F_d(\mathbf{x}_{k-1}) = F_d(F_d(\mathbf{x}_{k-2})) = \cdots := F_d^k(\mathbf{x}_0), \qquad (1.14)$$

where $k = 0, 1, 2, \ldots .$

Example 1.1. For the map

$$f(x) = p \, x(1-x), \qquad p \in R.$$

one has

$$f^2(x) = f(p \, x(1-x)) = p \, [p \, x(1-x)] \big(1 - [p \, x(1-x)]\big),$$

where the last equality is obtained by substituting each x in the previous step with $[px(1-x)]$. For a large number n of iterations, it quickly becomes very messy and actually impossible to write out the final explicit formula of the composite map $f^n(x)$.

If a function or map $\mathbf{f}$ is invertible, with inverse $\mathbf{f}^{-1}$, then one has $\mathbf{f}^{-2}(\mathbf{x}) = (\mathbf{f}^{-1})^2(\mathbf{x})$, and $\mathbf{f}^{-n}(\mathbf{x}) = (\mathbf{f}^{-1})^n(\mathbf{x})$, and so on. With the convention that $\mathbf{f}^0$ denotes the identity map, namely $\mathbf{f}^0(\mathbf{x}) := \mathbf{x}$, a general composition formula for an invertible $\mathbf{f}$ can be obtained:

$$\mathbf{f}^n(\mathbf{x}) = \mathbf{f} \circ \mathbf{f}^{n-1}(\mathbf{x}) = \mathbf{f}(\mathbf{f}^{n-1}(\mathbf{x})), \quad n = 0, \pm 1, \pm 2, \ldots . \qquad (1.15)$$

The derivative of a composite map can be obtained via the chain rule. For instance, in the 1-dimensional case,

$$(f^n)'(x_0) = f'(x_{n-1}) \cdots f'(x_0). \qquad (1.16)$$

This formula is convenient to use, because one does not need to explicitly compute $f^n(x)$, or $(f^n)'(x)$. Moreover, using the chain rule, one has

$$(f^{-1})' = \frac{1}{f'(x_{-1})}, \qquad x_{-1} := f^{-1}(x). \qquad (1.17)$$

Example 1.2. For $f(x) = x(1-x)$ with $x_0 = 1/2$ and $n = 3$, one has

$$f'(x) = 1 - 2x, \quad x_1 = f(x_0) = 1/4, \quad \text{and} \quad x_2 = f(x_1) = 3/16,$$

so that

$$\begin{aligned}
(f^3)'(1/2) &= f'(3/16)\, f'(1/4)\, f'(1/2) \\
&= \big(1 - 2(3/16)\big)\big(1 - 2(1/4)\big)\big(1 - 2(1/2)\big) \\
&= 0.
\end{aligned}$$

Finally, consider a function or map $\mathbf{f}$ given by either (1.13) or (1.14).

Definition 1.1. For a positive integer n, a point $\mathbf{x}^*$ is called a *periodic point of period n*, or an *n-periodic point*, of $\mathbf{f}$, if it satisfies

$$\mathbf{f}^n(\mathbf{x}^*) = \mathbf{x}^* \quad \text{but} \quad \mathbf{f}^k(\mathbf{x}^*) \neq \mathbf{x}^* \quad \text{for} \quad 0 < k < n. \tag{1.18}$$

If $\mathbf{x}^*$ is of period one ($n = 1$), then it is also called a *fixed point*, or an *equilibrium point*, which satisfies

$$\mathbf{f}(\mathbf{x}^*) = \mathbf{x}^*. \tag{1.19}$$

Moreover, a point $\mathbf{x}^*$ is said to be *eventually periodic of period n* if there is an integer $m > 0$ such that

$$\mathbf{f}^m(\mathbf{x}^*) \quad \text{is a periodic point} \quad \text{and} \quad \mathbf{f}^{m+n}(\mathbf{x}^*) = \mathbf{f}^m(\mathbf{x}^*). \tag{1.20}$$

Consequently,

$$\mathbf{f}^{n+q}(\mathbf{x}^*) = \mathbf{f}^q(\mathbf{x}^*) \quad \text{for all} \quad q \geq m.$$

This justifies the name "eventually".

Example 1.3. The map $f(x) = x^3 - x$ has three fixed points: $x_1^* = 0$ and $x_{1,2}^* = \pm\sqrt{2}$, which are solutions of the equation $f(x^*) = x^*$. It has two eventually fixed points of period one: $x_{1,2}^* = \pm 1$, since their first iterates go to the fixed point 0.

Definition 1.2. For a continuous-time function or map, $\mathbf{f}$, with a fixed point $\mathbf{x}^*$, the *forward orbit of $\mathbf{x}^*$* is

$$\Omega^+(\mathbf{x}^*) := \big\{ \mathbf{f}^k(\mathbf{x}^*): \ k \geq 0 \big\}.$$

If $\mathbf{f}$ is invertible, then the *backward orbit of $\mathbf{x}^*$* is

$$\Omega^-(\mathbf{x}^*) := \big\{ \mathbf{f}^k(\mathbf{x}^*): \ k \leq 0 \big\}.$$

The *whole orbit of $\mathbf{x}^*$*, thus, is

$$\Omega(\mathbf{x}^*) = \Omega^+(\mathbf{x}^*) \cup \Omega^-(\mathbf{x}^*) = \big\{ \mathbf{f}^k(\mathbf{x}^*): \ k = 0, \pm 1, \pm 2, \dots \big\}.$$

Definition 1.3. For a continuous-time function or map, $\mathbf{f}$, a set $\mathcal{S} \subset R^n$ is said to be *forward invariant* under $\mathbf{f}$, if $\mathbf{f}^k(\mathbf{x}) \in \mathcal{S}$ for all $\mathbf{x} \in \mathcal{S}$ and for all $k = 0, 1, 2, \ldots$. Furthermore, for an invertible $\mathbf{f}$, a set $\mathcal{S} \subset R^n$ is said to be *backward invariant* under $\mathbf{f}$, if $\mathbf{f}^k(\mathbf{x}) \in \mathcal{S}$ for all $\mathbf{x} \in \mathcal{S}$ and for all $k = 0, -1, -2, \ldots$.

1.3 Dynamical Analysis on Phase Planes

In this section, a general 2-dimensional nonlinear autonomous system is considered:

$$\begin{aligned} \dot{x} &= f(x, y), \\ \dot{y} &= g(x, y). \end{aligned} \tag{1.21}$$

In this system, the two functions f and g together describe the *vector field* of the system. Here and in the following, for 2- or 3-dimensional systems, the state variables will be denoted as x, y, and z, instead of x_1, x_2, and x_3, for notational convenience.

1.3.1 *Phase Plane of a Planar System*

The path traveled by a solution of the continuous-time planar system (1.21), starting from an initial state (x_0, y_0), is a *solution trajectory* or *orbit* of the system, and is sometimes denoted as $\varphi_t(x_0, y_0)$.

For autonomous systems, the x–$\dot{x}$ coordinate plane is called the *phase plane* of the system. In general, even if $y \neq \dot{x}$, the x–y coordinate plane is called the (*generalized*) *phase plane*. In the higher-dimensional case, it is called the *phase space* of the underlying dynamical system. Moreover, the orbit family of an autonomous system, corresponding to all possible initial conditions, is called a *solution flow* in the phase space. The graphical layout of the solution flow provides a *phase portrait* of the system dynamics in the phase space, as depicted by Fig. 1.2.

The phase portrait of the damped and undamped pendulum systems, (1.1) and (1.2), are shown in Fig. 1.3.

Examining the phase portraits shown in Fig. 1.3, a natural question arises: how can one determine the motion direction of the orbit flow in the phase plane as the time evolves? Clearly, a computer graphic demonstration can provide a fairly complete answer to this question. However, a quick sketch to show the *qualitative behavior* of the system dynamics is still quite possible, as illustrated by the following two examples.

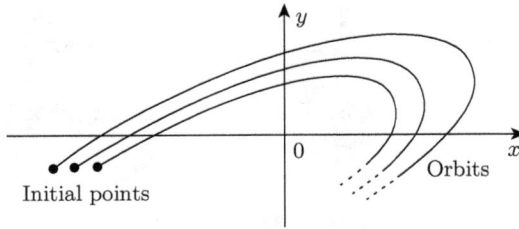

Fig. 1.2   Phase portrait on the phase plane of a dynamical system.

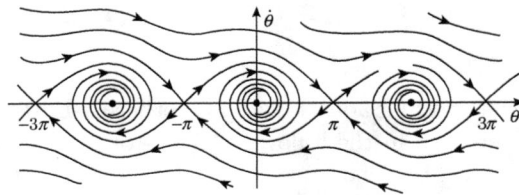

(a) damped pendulum

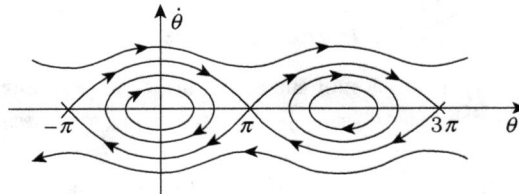

(b) undamped pendulum

Fig. 1.3   Phase portraits of the damped and undamped pendula.

Example 1.4. Consider the simple linear harmonic oscillator

$$\ddot{\theta} + \theta = 0\,.$$

By defining

$$x = \theta \qquad \text{and} \qquad y = \dot{\theta}\,,$$

this harmonic equation becomes

$$\dot{x} = y\,,$$
$$\dot{y} = -x\,.$$

With initial conditions $x(0) = 1$ and $y(0) = 0$, this equation has solution

$$x(t) = \cos(t) \quad \text{and} \quad y(t) = -\sin(t).$$

The solution trajectory in the x–y–t space and the corresponding orbit on the x–y phase plane are sketched in Fig. 1.4, together with some other solutions starting from different initial conditions. This shows clearly the direction of motion of the phase portrait.

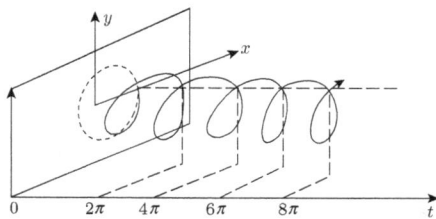

(a) phase portrait in the x–y–t space

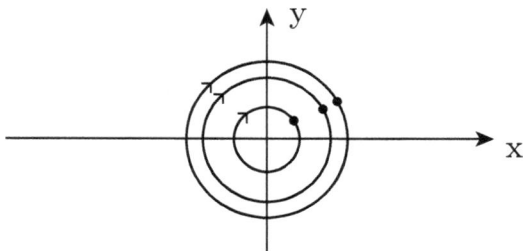

(b) phase portrait on the x–y phase plane

Fig. 1.4   Phase portraits of the simple harmonic equation.

Example 1.5. Consider the normalized undamped pendulum equation

$$\ddot{\theta} + \sin(\theta) = 0.$$

By defining

$$x = \theta \quad \text{and} \quad y = \dot{\theta},$$

this pendulum equation can be written as

$$\dot{x} = y,$$
$$\dot{y} = -\sin(x).$$

With initial conditions $x(0) = 1$ and $y(0) = 0$, it has solution

$$x(t) = \theta(t) = 2 \sin^{-1}\left(\tanh(t)\right) \qquad \text{and} \qquad y(t) = \dot{\theta}(t).$$

The phase portrait of this undamped pendulum, along with some other solutions starting from different initial conditions, is sketched on the x–y phase plane shown in Fig. 1.5. This sketch also clearly indicates the direction of motion of the solution flow. The shape of a solution trajectory of this undamped pendulum in the x–y–t space can also be sketched, which could be quite complex however, depending on the initial conditions.

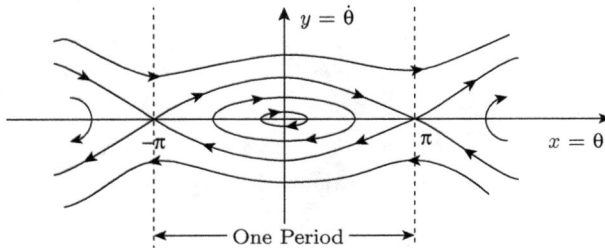

Fig. 1.5   Phase portrait of the undamped pendulum equation.

Example 1.6. Another way to understand the phase portrait of the general undamped pendulum

$$\dot{x} = y,$$
$$\dot{y} = -\frac{g}{\ell}\sin(x)$$

is to examine its total (kinetic and potential) energy

$$E = \frac{y^2}{2} + \frac{g}{\ell}\int_0^x \sin(\sigma)\, d\sigma = \frac{y^2}{2} + \frac{g}{\ell}\left(1 - \cos(x)\right).$$

Figure 1.6 shows the potential energy plot, $P(x)$, versus $x = \theta$, and the corresponding phase portrait on the x–y phase plane of the damped pendulum. It is clear that the lowest level of total energy is $E = 0$, which corresponds to the angular positions $x = \theta = \pm 2n\pi$, $n = 0, 1, \ldots$. As the total energy increases, the pendulum swings up or down, with an increasing or decreasing angular speed, $|y| = |\dot{\theta}|$, provided that E is within its limit indicated by E_2. Within each period of oscillation, the total energy $E = $ constant, according to the conservation law of energy, for this idealized undamped pendulum.

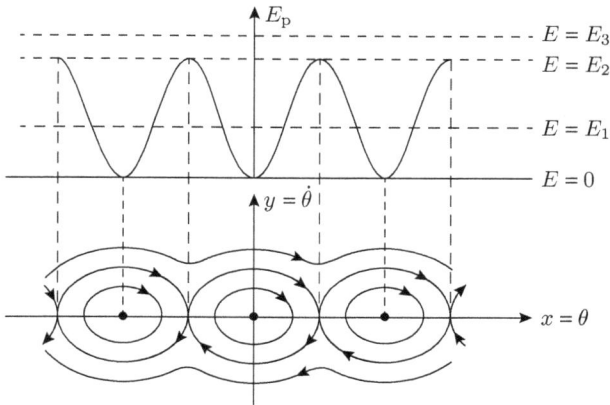

Fig. 1.6   Phase portrait of the undamped pendulum versus its total energy.

1.3.2   *Analysis on Phase Planes*

This subsection addresses the following question: why is it important to study autonomous systems and their phase portraits?

The answer to this question is provided by the following several theorems, which together summarize a few important and useful properties of autonomous systems in the study of nonlinear dynamics. Although these theorems are stated and proven for planar systems in this subsection, they generally hold for higher-dimensional autonomous systems as well.

Theorem 1.1. *A nonautonomous system can be equivalently reformulated as an autonomous one.*

Proof. Consider a general nonautonomous system,

$$\dot{\mathbf{x}} = \mathbf{f}(\mathbf{x}, t), \qquad \mathbf{x}_0 \in R^n.$$

Let the independent time variable t be a new variable by defining $x_{n+1}(t) = t$ for this separated variable t of the system. Then, $\dot{x}_{n+1} = 1$. Consequently, the original system can be equivalently reformulated by augmenting it as

$$\begin{bmatrix} \dot{\mathbf{x}} \\ \dot{x}_{n+1} \end{bmatrix} = \begin{bmatrix} \mathbf{f}(\mathbf{x}, x_{n+1}) \\ 1 \end{bmatrix},$$

which is an autonomous system.   □

Obviously the price to pay for this conversion, from a nonautonomous system to an autonomous one, is the increase of dimension. In dynamical systems analysis, this usually is acceptable since the increase is only by one, which is not a big deal for higher-dimensional systems. Nevertheless, this shows that, without loss of generality, one may only discuss autonomous systems in nonlinear dynamical analysis especially in higher-dimensional cases.

However, it is important to note that, in a nonlinear control system of the form

$$\dot{\mathbf{x}} = \mathbf{f}\big(\mathbf{x}, \mathbf{u}(t)\big),$$

which will be studied in detail later in the book, the controller $\mathbf{u}(t)$ is a time function and is yet to be designed, which it is not a system variable. In this case, one should not (cannot) convert the control system to be autonomous using this technique; otherwise, the controller loses its physical meaning and can never be designed for the intended control tasks. This issue will be revisited later within the context of feedback controllers design.

Theorem 1.2. *If $\mathbf{x}(t)$ is a solution of the autonomous system $\dot{\mathbf{x}} = \mathbf{f}(\mathbf{x})$, then so is the trajectory $\mathbf{x}(t + a)$, for any real constant a. Moreover, these two solutions are the same, except that they may pass the same point on the phase plane at two different time instants.*

The last statement of the theorem describes the inherent time-invariant property of autonomous systems.

Proof. Because $\frac{d}{dt}\mathbf{x}(t) = \mathbf{f}(\mathbf{x}(t))$, for any real constant τ, one has

$$\frac{d}{dt}\mathbf{x}(t+a)\bigg|_{t=\tau} = \frac{d}{ds}\mathbf{x}(s)\bigg|_{s=\tau+a} = \mathbf{f}(\mathbf{x}(s))\big|_{s=\tau+a} = \mathbf{f}(\mathbf{x}(t+a))\big|_{t=\tau}.$$

Since this holds for all real τ, it implies that $\mathbf{x}(t + a)$ is a solution of the equation $\dot{\mathbf{x}} = \mathbf{f}(\mathbf{x})$. Moreover, the value assumed by $\mathbf{x}(t)$ at time instant $t = t^*$ is the same as that assumed by $\mathbf{x}(t + a)$ at time instant $t = t^* - a$. Hence, these two solutions are identical, in the sense that they have the same trajectory if they are both plotted on the same phase plane.   □

Example 1.7. The autonomous system $\dot{x}(t) = x(t)$ has a solution $x(t) = e^t$. It is easy to verify that e^{t+a} is also a solution of this system for any real constant a. These two solutions are the same, in the sense that they have the same trajectory if they are plotted on the x–$\dot{x}$ phase plane, except that they pass the same point at two different time instants; for instance, the first one passes the point $(x, \dot{x}) = (1, 1)$ at $t = 0$ but the second, at $t = -a$.

However, a nonautonomous system may not have such a property.

Example 1.8. The nonautonomous system $\dot{x}(t) = e^t$ has a solution $x(t) = e^t$. But e^{t+a} is not its solution if $a \neq 0$.

Note that, if one applies Theorem 1.1 to Example 1.8 and let $y(t) = t$, then

$$\dot{x}(t) = e^{y(t)},$$
$$\dot{y}(t) = 1,$$

(a)

which has solution

$$x(t) = e^t,$$
$$y(t) = t.$$

(b)

Theorem 1.1 states that (b) is a solution of (a), which does not mean that (b) is a solution of the original equation $\dot{x}(t) = e^t$. In fact, only the first part of (b), i.e. $x(t) = e^t$, is a solution of the original equation, and the second part of (b) is merely used to convert the given nonautonomous system to be an autonomous one.

Theorem 1.3. *Suppose that a given autonomous system $\dot{\mathbf{x}}(t) = \mathbf{f}(\mathbf{x}(t))$ has a unique solution starting from an initial state $\mathbf{x}(t_0) = \mathbf{x}_0$. Then, there will not be any other (different) orbit of the same system that also passes through this same point $\mathbf{x}_0$ on the phase plane at any time.*

Before giving a proof to this result, two remarks are in order. First, this theorem implies that the solution flow of an autonomous system has simple geometry, as depicted in Fig. 1.7, where different orbits starting from different initial states do not cross each other. Second, in the phase portrait of the (damped or undamped) pendulum (see Fig. 1.3, it may seem that there are more than one orbit passing through the points $(-\pi, 0)$ and $(\pi, 0)$ etc. However, those orbits are periodic orbits, so the principal solution of the pendulum corresponds to those curves located between the two vertical lines passing through the two end points $x = -\pi$ and $x = \pi$, respectively. Thus, within each 2π period, actually no self-crossing exists. It will be seen later that all such seemingly self-crossing occur only at those special points called *stable node (sink)*, *unstable node (source)*, or *saddle node* (see Fig. 1.8), where the orbits either spiral into a sink, spiral out from a source, or spiral in and out from a saddle node in different directions.

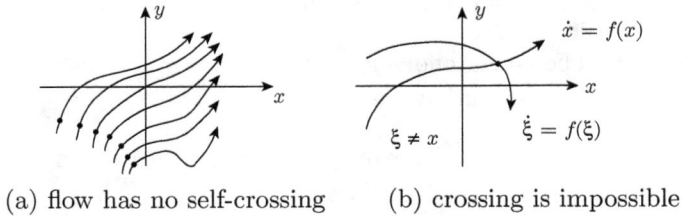

(a) flow has no self-crossing   (b) crossing is impossible

Fig. 1.7   Simple phase portrait of an autonomous system.

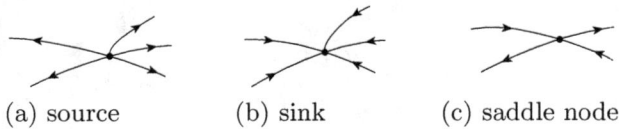

(a) source   (b) sink   (c) saddle node

Fig. 1.8   Simple phase portrait of an autonomous system.

Proof. Let $\mathbf{x}_1$ and $\mathbf{x}_2$ be two solutions of $\dot{\mathbf{x}} = \mathbf{f}(\mathbf{x})$, satisfying

$$\mathbf{x}_1(t_1) = \mathbf{x}_0 \quad \text{and} \quad \mathbf{x}_2(t_2) = \mathbf{x}_0, \quad t_1 \neq t_2.$$

By Theorem 1.2, one has

$$\widetilde{\mathbf{x}}_2(t) := \mathbf{x}_2\big(t - (t_1 - t_2)\big),$$

which is the same solution of the given autonomous system, namely,

$$\mathbf{x}_2(t) = \widetilde{\mathbf{x}}_2(t). \tag{a}$$

This solution satisfies

$$\widetilde{\mathbf{x}}_2(t_1) := \mathbf{x}_2\big(t_1 - (t_1 - t_2)\big) = \mathbf{x}_2(t_2) = \mathbf{x}_0.$$

Therefore, by the uniqueness of the solution, $\mathbf{x}_1$ and $\widetilde{\mathbf{x}}_2$ are the same:

$$\mathbf{x}_1(t) = \widetilde{\mathbf{x}}_2(t), \tag{b}$$

since they are both equal to $\mathbf{x}_0$ at the same initial time t_1. Thus, (a) and (b) together imply that $\mathbf{x}_1$ and $\mathbf{x}_2$ are identical. $\square$

Note that a nonautonomous system may not have such a property.

Example 1.9. Consider the nonautonomous system

$$\dot{x}(t) = \cos(t).$$

This system has the following solutions, among others:

$$x_1(t) = \sin(t) \qquad \text{and} \qquad x_2(t) = 1 + \sin(t).$$

These two solutions are different, for if they are plotted on the phase plane, they show two different trajectories:

$$\dot{x}_1(t) = \cos(t) = \pm\sqrt{1 - \sin^2(t)} = \pm\sqrt{1 - x_1^2},$$

$$\dot{x}_2(t) = \cos(t) = \pm\sqrt{1 - \sin^2(t)} = \pm\sqrt{1 - [1 + \sin(t) - 1]^2}$$
$$= \pm\sqrt{1 - (x_2 - 1)^2}.$$

These two trajectories cross over at a point, $(x_1, x_2) = (1/2, 1/2)$, as can be seen from Fig. 1.9.

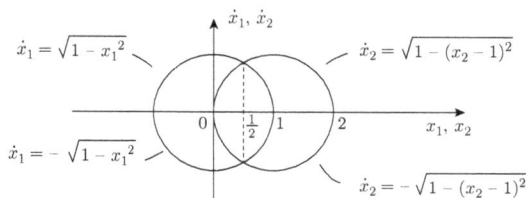

Fig. 1.9   Two crossing trajectories of a nonautonomous system.

Theorem 1.4. *A closed orbit of the autonomous system* $\dot{\mathbf{x}} = \mathbf{f}(\mathbf{x})$ *on the phase plane corresponds to a periodic solution of the system.*

Proof. For a τ-periodic solution, $\mathbf{x}(t)$, one has $\mathbf{x}(t_0 + \tau) = \mathbf{x}(t_0)$ for any $t_0 \in R$, which means that the trajectory of $\mathbf{x}(t)$ is closed.

On the contrary, suppose that the orbit of $\mathbf{x}(t)$ is closed. Let $\mathbf{x}_0$ be a point in the closed orbit. Then, $\mathbf{x}_0 = \mathbf{x}(t_0)$ for some t_0, and the trajectory of $\mathbf{x}(t)$ will return to $\mathbf{x}_0$ after some time, say $\tau \geq 0$; that is, $\mathbf{x}(t_0 + \tau) = \mathbf{x}_0 = \mathbf{x}(t_0)$. Since $\mathbf{x}_0$ is arbitrary, and so is t_0, this implies that $\mathbf{x}(t + \tau) = \mathbf{x}(t)$ for all t, meaning that $\mathbf{x}(t)$ is periodic with period τ. $\qquad\square$

Yet, a nonautonomous system may not have such a property.

Example 1.10. The nonautonomous system

$$\dot{x} = 2ty,$$
$$\dot{y} = -2tx,$$

has solution

$$x(t) = \alpha \, \cos(t^2) + \beta \, \sin(t^2) \,,$$
$$y(t) = -\alpha \, \sin(t^2) + \beta \, \cos(t^2) \,,$$

for some constants α and β determined by initial conditions. This solution is not periodic, but it is a circle (a closed orbit) on the x–y phase plane.

As mentioned at the beginning of this subsection, the above four theorems hold for general higher-dimensional autonomous systems. Since these properties are simple, elegant and easy to use, which a nonautonomous system may not have, it is very natural to focus a general study of complex dynamics on autonomous systems in various forms with any dimensions. This motivates the following investigations.

1.4   Qualitative Behaviors of Dynamical Systems

In this section, consider a general 2-dimensional autonomous system,

$$\dot{x} = f(x, y) \,,$$
$$\dot{y} = g(x, y) \,. \tag{1.22}$$

Let Γ be a periodic solution of the system which, as discussed above, has a closed orbit on the x–y phase plane.

Definition 1.4. Γ is said to be an *inner* (*outer*) *limit cycle* of system (1.22) if, in an arbitrarily small neighborhood of the inner (*outer*) region of Γ, there is always (part of) a nonperiodic solution orbit of the system. Γ is called a *limit cycle*, if it is both inner and outer limit cycles.

Simply put, a limit cycle is a periodic orbit of the system that corresponds to a closed orbit on the phase plane and possesses certain (attracting or repelling) limiting properties. Figure 1.10 shows some typical limit cycles for the 2-dimensional system (1.22), where the attracting limit cycle is said to be *stable*, while the repelling one, *unstable*.

Example 1.11. The simple harmonic oscillator discussed in Example 1.4 has no limit cycles. The solution flow of the system constitutes a ring of periodic orbits, called *periodic ring*, as shown in Fig. 1.4. Similarly, the undamped pendulum has no limit cycles, as shown in Fig. 1.3.

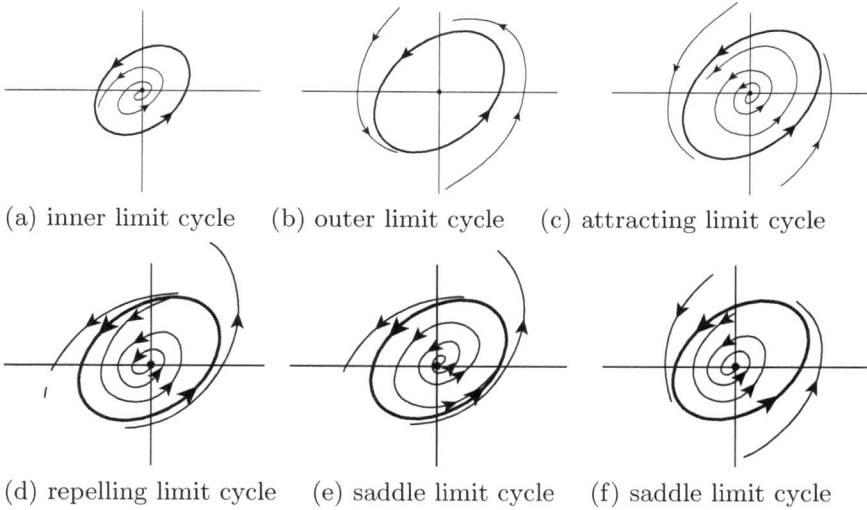

(a) inner limit cycle   (b) outer limit cycle   (c) attracting limit cycle

(d) repelling limit cycle   (e) saddle limit cycle   (f) saddle limit cycle

Fig. 1.10   Periodic orbits and limit cycles.

This example shows that, although a limit cycle is a periodic orbit, not all periodic orbits are limit cycles, not even inner or outer limit cycles.

Example 1.12. A typical example of a stable limit cycle is the periodic solution of the Rayleigh oscillator, described by

$$\ddot{x} + x = p\left(\dot{x} - \dot{x}^3\right), \qquad p > 0, \tag{1.23}$$

which was formulated in the 1920s to describe oscillations in some electrical and mechanical systems. This limit cycle is shown in Fig. 1.11 for some different values of p. These phase portraits are usually obtained either numerically or experimentally, because they do not have simple analytic formulas.

Example 1.13. Another typical example of a stable limit cycle is the periodic solution of the van der Pol oscillator, described by

$$\ddot{x} + x = p\left(1 - x^2\right)\dot{x}, \qquad p > 0, \tag{1.24}$$

which was formulated around 1920 to describe oscillations in a triode circuit. This limit cycle is shown in Fig. 1.12, which is usually obtained either numerically or experimentally, because it does not have a simple analytic formula.

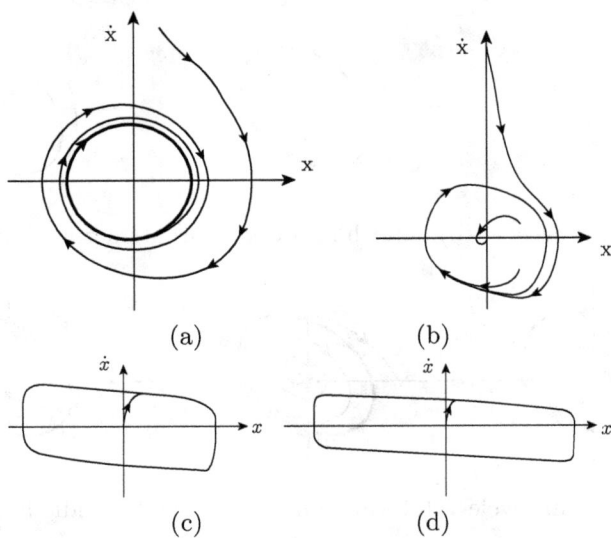

Fig. 1.11   Phase portrait of the Rayleigh oscillator. (a) $p = 0.01$; (b) $p = 0.1$; (c) $p = 1.0$; (d) $p = 10.0$.

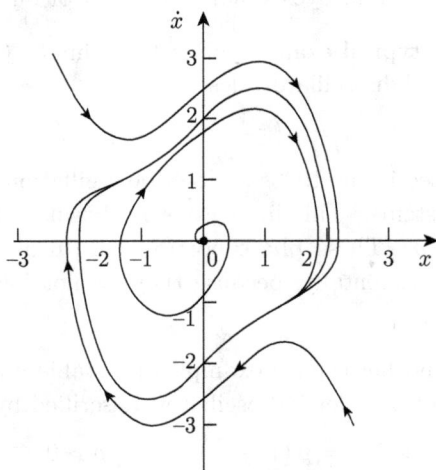

Fig. 1.12   Phase portrait of the van der Pol oscillator.

1.4.1 Qualitative Analysis of Linear Dynamics

For illustration, consider a 2-dimensional linear autonomous (i.e. time-invariant) system,

$$\dot{\mathbf{x}}(t) = A\mathbf{x}(t), \qquad \mathbf{x}(0) = \mathbf{x}_0, \tag{1.25}$$

where A is a given 2×2 constant matrix and, for simplicity, the initial time is set to $t_0 = 0$.

Obviously, this system has a unique fixed point $\mathbf{x}^* = 0$ and has a unique solution $\mathbf{x}(t) = e^{tA}\mathbf{x}_0$. Decompose its solution as

$$\mathbf{x}(t) = e^{tA}\mathbf{x}_0 = M e^{tJ} M^{-1}\mathbf{x}_0, \tag{1.26}$$

where $M = \begin{bmatrix} \mathbf{v}_1 & \mathbf{v}_2 \end{bmatrix}$ with $\mathbf{v}_1$ and $\mathbf{v}_2$ being two linearly independent real eigenvectors associated with the two eigenvalues of A, and J is in the Jordan canonical form, which is one of the following three possible forms:

$$\begin{bmatrix} \lambda_1 & 0 \\ 0 & \lambda_2 \end{bmatrix}, \qquad \begin{bmatrix} \lambda & \kappa \\ 0 & \lambda \end{bmatrix}, \qquad \begin{bmatrix} \alpha & -\beta \\ \beta & \alpha \end{bmatrix},$$

with $\lambda_1, \lambda_2, \lambda, \alpha$, and β being real constants, and $\kappa = 0$ or 1. Note that for the third case, its eigenvalues are complex conjugates: $\lambda_{1,2} = \alpha \pm j\beta$, where $j = \sqrt{-1}$.

Thus, there are three cases to study, according to the three different canonical forms of the Jordan matrix J shown above.

Case (i). The two constants λ_1 and λ_2 are different, but both real and nonzero.

In this case, λ_1 and λ_2 are the eigenvalues of matrix A, associated with two eigenvectors $\mathbf{v}_1$ and $\mathbf{v}_2$, respectively. Let

$$\mathbf{z} = M^{-1}\mathbf{x},$$

where $M = \begin{bmatrix} \mathbf{v}_1 & \mathbf{v}_2 \end{bmatrix}$. Then, the given system is transformed to

$$\dot{\mathbf{z}} = \begin{bmatrix} \lambda_1 & 0 \\ 0 & \lambda_2 \end{bmatrix} \mathbf{z}, \qquad \text{with} \quad \mathbf{z}_0 = M^{-1}\mathbf{x}_0 := \begin{bmatrix} z_{10} \\ z_{20} \end{bmatrix}.$$

Its solution is

$$z_1(t) = z_{10}\, e^{t\lambda_1} \qquad \text{and} \qquad z_2(t) = z_{20}\, e^{t\lambda_2},$$

which are related by

$$z_2(t) = c\, z_1^{\lambda_2/\lambda_1}(t), \qquad \text{with} \quad c = z_{20}\left(z_{10}\right)^{-\lambda_2/\lambda_1}.$$

To show the phase portraits of the solution flow, there are three situations to consider: (a) $\lambda_2 < \lambda_1 < 0$; (b) $\lambda_2 < 0 < \lambda_1$; (c) $0 < \lambda_2 < \lambda_1$.

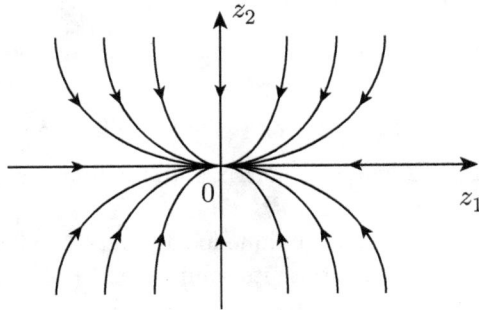

Fig. 1.13   Phase portrait of the transformed case (a): $\lambda_2 < \lambda_1 < 0$.

Only situation (a) is discussed in detail here. In this case, the two eigenvalues are both negative, so that $e^{t\lambda_1} \to 0$ and $e^{t\lambda_2} \to 0$ as $t \to \infty$, but the latter tends to zero faster. The corresponding phase portrait is shown in Fig. 1.13, where the fixed point (the origin) is a stable node.

Now, return to the original state, $\mathbf{x} = M\mathbf{z}$. The original phase portrait is somewhat twisted, as shown in Fig. 1.14. Figures 1.13 and 1.14 are topologically equivalent, hence can be considered to be the same qualitatively. A more precise meaning of topological equivalence will be given later in (1.29). Roughly speaking, it means that their dynamical behaviors are qualitatively similar.

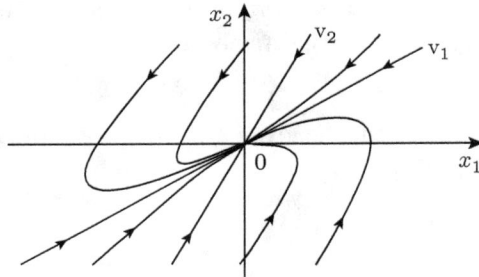

Fig. 1.14   Phase portrait of the original case (a): $\lambda_2 < \lambda_1 < 0$.

The other two situations, (b) and (c), can be analyzed in the same way, where case (b) shows a saddle node and case (c), an unstable node. This is left as an exercise to sketch.

Case (ii). The two constants λ_1 and λ_2 are nonzero complex conjugates: $\lambda_{1,2} = \alpha \pm j\,\beta$, where $j = \sqrt{-1}$. Let

$$\mathbf{z} = M^{-1}\mathbf{x}\,,$$

and transform the given system to

$$\dot{\mathbf{z}} = \begin{bmatrix} \alpha & -\beta \\ \beta & \alpha \end{bmatrix} \mathbf{z}\,, \qquad \text{with} \quad \mathbf{z}_0 = \begin{bmatrix} z_{10} \\ z_{20} \end{bmatrix}.$$

In polar coordinates,

$$r = \sqrt{z_1^2 + z_2^2} \qquad \text{and} \qquad \theta = \tan^{-1}\left(\frac{z_2}{z_1}\right),$$

which has solution

$$r(t) = r_0\,e^{\alpha t} \qquad \text{and} \qquad \theta(t) = \theta_0 + \beta\,t\,,$$

where $r_0 = (z_{10}^2 + z_{20}^2)^{1/2}$ and $\theta_0 = \tan^{-1}(z_{20}/z_{10})$. This solution trajectory is visualized by Fig. 1.15, where the fixed point (the origin) in case (a) is called a *stable node*, in case (b), an *unstable focus*, and in case (c), a *center*. On the original x–y phase plane, the phase portrait has a twisted shape, as shown in Fig. 1.16.

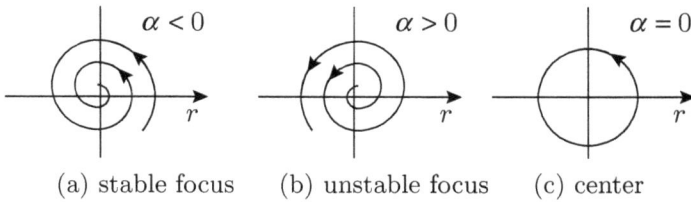

(a) stable focus   (b) unstable focus   (c) center

Fig. 1.15   Phase portrait of the transformed Case (ii): $\lambda_{1,2} = \alpha \pm j\,\beta$.

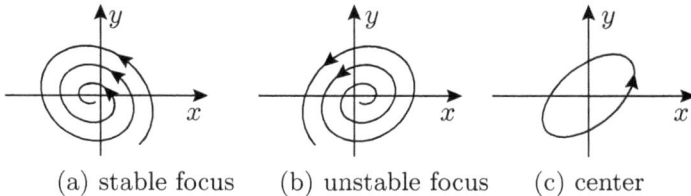

(a) stable focus   (b) unstable focus   (c) center

Fig. 1.16   Phase portrait of the original Case (ii): $\lambda_{1,2} = \alpha \pm j\,\beta$.

Case (iii). The two constants λ_1 and λ_2 are nonzero multiple real values: $\lambda_1 = \lambda_2 := \lambda$. Let

$$\mathbf{z} = M^{-1}\mathbf{x},$$

and transform the given system to

$$\dot{\mathbf{z}} = \begin{bmatrix} \lambda & \kappa \\ 0 & \lambda \end{bmatrix} \mathbf{z}, \quad \text{with} \quad \mathbf{z}_0 = \begin{bmatrix} z_{10} \\ z_{20} \end{bmatrix}.$$

Its solution is

$$z_1(t) = e^{\lambda t}\left(z_{10} + \kappa\, z_{20} t \right) \quad \text{and} \quad z_2(t) = z_{20}\, e^{\lambda t},$$

which are related via

$$z_1(t) = z_2(t)\left[\frac{z_{10}}{z_{20}} + \frac{\kappa}{\lambda} \ln\left(\frac{z_2(t)}{z_{20}} \right) \right].$$

Its phase portrait is shown in Fig. 1.17, and its corresponding phase portrait on the x–y phase plane is similar; in particular, the linear coordinates that transform M do not change the shape of any straight line on the two phase planes.

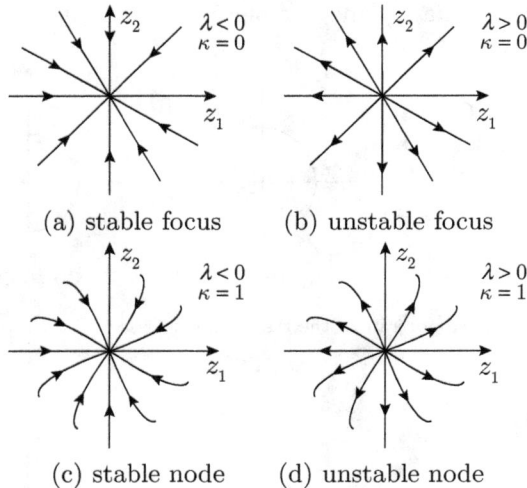

(a) stable focus   (b) unstable focus

(c) stable node   (d) unstable node

Fig. 1.17   Phase portrait of Case (iii): $\lambda_1 = \lambda_2 \neq 0$.

Case (iv). One, or both, of $\lambda_{1,2}$ is zero.

In this degenerate case, the matrix A in $\dot{\mathbf{x}} = A\mathbf{x}$ has a nontrivial null space, of dimension 1 or 2 respectively, so that any vector in the null space of A is a fixed point. As a result, the system has a *fixed* or *equilibrium subspace*. Specifically, these two situations are as follows:

(a) $\lambda_1 = 0$ but $\lambda_2 \neq 0$

In this case, the system can be transformed to

$$\dot{\mathbf{z}} = \begin{bmatrix} 0 & 0 \\ 0 & \lambda_2 \end{bmatrix} \mathbf{z}, \quad \text{with} \quad \mathbf{z}_0 = \begin{bmatrix} z_{10} \\ z_{20} \end{bmatrix},$$

which has solution

$$z_1(t) = z_{10} \quad \text{and} \quad z_2(t) = z_{20}\, e^{\lambda_2 t}.$$

The phase portrait of $\mathbf{x} = M\mathbf{z}$ is shown in Fig. 1.18, where (a) shows a *stable equilibrium subspace* and (b), an *unstable subspace*.

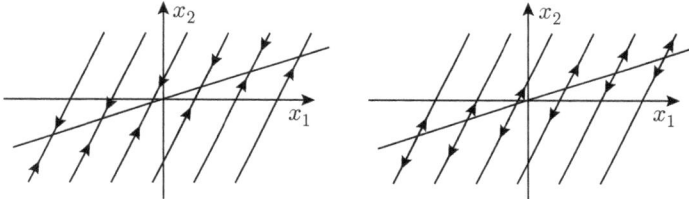

(a) a stable equilibrium subspace   (b) an unstable equilibrium subspace

Fig. 1.18   Phase portrait of Case (iv) (a): $\lambda_1 = 0$ but $\lambda_2 \neq 0$.

(b) $\lambda_1 = \lambda_2 = 0$

In this case, the system is transformed to

$$\dot{\mathbf{z}} = \begin{bmatrix} 0 & 1 \\ 0 & 0 \end{bmatrix} \mathbf{z}, \quad \text{with} \quad \mathbf{z}_0 = \begin{bmatrix} z_{10} \\ z_{20} \end{bmatrix},$$

which has solution

$$z_1(t) = z_{10} + z_{20}\, t \quad \text{and} \quad z_2(t) = z_{20}.$$

The phase portrait of $\mathbf{x} = M\mathbf{z}$ is shown in Fig. 1.19, which is a *saddle equilibrium subspace*.

1.4.2   Qualitative Analysis of Nonlinear Dynamics

This subsection is devoted to some qualitative analysis of dynamical behaviors of a general nonlinear autonomous system in a neighborhood of a

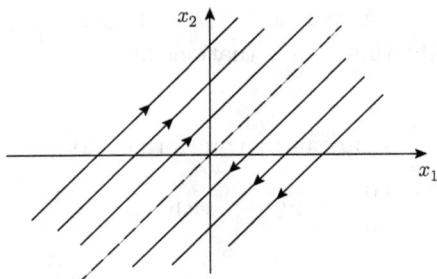

Fig. 1.19   Phase portrait of Case (iv) (b): $\lambda_1 = \lambda_2 = 0$.

fixed point (equilibrium point) of the system. Therefore, unlike the linear systems discussed above, all results derived here are *local*.

Consider a general nonlinear autonomous system,

$$\dot{\mathbf{x}} = \mathbf{f}(\mathbf{x}), \qquad \mathbf{x}_0 \in R^n, \tag{1.27}$$

where it is assumed that $\mathbf{f} \in C^1$, i.e. it is continuously differentiable with respect to its arguments, and that the system has a fixed point, $\mathbf{x}^*$.

Taylor-expanding $\mathbf{f}(\mathbf{x})$ at $\mathbf{x}^*$ yields

$$\dot{\mathbf{x}} = \mathbf{f}(\mathbf{x}^*) + \left[\frac{\partial \mathbf{f}}{\partial \mathbf{x}}\right]_{\mathbf{x}=\mathbf{x}^*} (\mathbf{x} - \mathbf{x}^*) + \mathbf{e}(\mathbf{x}) = J(\mathbf{x} - \mathbf{x}^*) + \mathbf{e}(\mathbf{x}),$$

where

$$J = \left[\frac{\partial \mathbf{f}}{\partial \mathbf{x}}\right]_{\mathbf{x}=\mathbf{x}^*} = \begin{bmatrix} \partial f_1/\partial x_1 & \cdots & \partial f_1/\partial x_n \\ \vdots & & \vdots \\ \partial f_n/\partial x_1 & \cdots & \partial f_n/\partial x_n \end{bmatrix}_{\mathbf{x}=\mathbf{x}^*}$$

is the Jacobian, and $\mathbf{e}(\mathbf{x}) = o(||\mathbf{x}||)$ represents the residual of all higher-order terms, which satisfies

$$\lim_{||\mathbf{x}|| \to \infty} \frac{||\mathbf{e}(\mathbf{x})||}{||\mathbf{x}||} = 0.$$

Letting $\mathbf{y} = \mathbf{x} - \mathbf{x}^*$ leads to

$$\dot{\mathbf{y}} = J\mathbf{y} + \mathbf{e}(\mathbf{y}),$$

where $\mathbf{e}(\mathbf{y}) = o(||\mathbf{y}||)$. In a small neighborhood of $\mathbf{x}^*$, $||\mathbf{x} - \mathbf{x}^*||$ is small, so $o(||\mathbf{y}||) \approx 0$. Thus, the nonlinear autonomous system (1.27) and its linearized system $\dot{\mathbf{x}} = J(\mathbf{x} - \mathbf{x}^*)$ have the same dynamical behaviors; particularly, the latter in a small neighborhood of $\mathbf{x}^*$ is the same as $\dot{\mathbf{y}} = J\mathbf{y}$ in a small neighborhood of 0. In other words, between $\mathbf{x}$ and $\mathbf{y}$, the following are comparable:

$$\dot{\mathbf{x}} = \mathbf{f}(\mathbf{x}) \qquad \text{versus} \qquad \dot{\mathbf{y}} = J\,\mathbf{y} \qquad (1.28)$$

$$\mathbf{x}^* \text{ is } \begin{cases} \text{stable node} \\ \text{unstable node} \\ \text{stable focus} \\ \text{unstable focus} \\ \text{saddle node} \end{cases} \iff \mathbf{y} = 0 \text{ is } \begin{cases} \text{stable node} \\ \text{unstable node} \\ \text{stable focus} \\ \text{unstable focus} \\ \text{saddle node} \end{cases} \qquad (1.29)$$

In this sense, the local dynamical behaviors of the two systems in (1.28) are said to be *qualitatively the same,* or *topologically equivalent.* A precise mathematical definition is given as follows.

Definition 1.5. Two time-invariant system functions, $f : X \to Y$ and $g : X^* \to Y^*$, where X, Y, X^*, and Y^* are (open sets of) metric spaces, are said to be *topologically equivalent,* if there is a homeomorphism, $h : Y \to Y^*$, such that $h^{-1} : X^* \to X$ and

$$g(x) = h^{-1} \circ f \circ h(x), \qquad x \in X,$$

where $\circ$ is the composite operation of two maps.

This definition is illustrated in Fig. 1.20. Here, a *homeomorphism* is an invertible continuous function whose inverse is also continuous. For instance, for $X = Y = R$, the two functions $f(x) = 2x^3$ and $g(x) = 8x^3$ are topologically equivalent. This is because one can find a homeomorphisim, $h(x) = (x)^{1/3}$, which yields

$$h^{-1} \circ f \circ h(x) = \left(2\left((x)^{1/3} \right)^3 \right)^3 = 8x^3 = g(x).$$

A homeomorphism preserves the system dynamics as seen from the one-one correspondence (1.29). When both X and Y are Euclidean spaces, the homeomorphism h may be viewed as a nonsingular coordinates transform.

For discrete-time systems (maps), such topological equivalence is also called the *topological conjugacy,* and the two maps are said to be *topologically conjugate* if they satisfy the relationships shown in Fig. 1.20, where h is a homeomorphism.

Theorem 1.5. *If f and g are topologically conjugate, then*

(i) *the orbits of f are mapped to the orbits of g under h;*
(ii) *if x^* is a fixed point of f, then the eigenvalues of $f'(x^*)$ are mapped to the eigenvalues of $g'(x^*)$.*

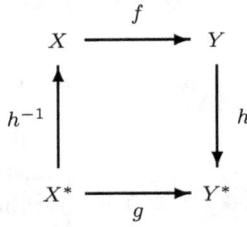

Fig. 1.20   Two topologically equivalent functions or topologically conjugate maps.

Proof. First, note that the orbit of x^* under iterates of map f is

$$\Omega(x^*) = \{\ \ldots, f^{-k}(x^*), \ldots, f^{-1}(x^*), x^*, f(x^*), \ldots, f^k(x^*)\ \}\,.$$

Since $f = h^{-1} \circ g \circ h$, for any given $k > 0$ one has

$$f^k(x^*) = (h^{-1} \circ g \circ h) \circ \cdots \circ (h^{-1} \circ g \circ h)(x^*)$$
$$= h^{-1} \circ g^k \circ h(x^*)\,.$$

On the other hand, since $f^{-1} = h^{-1} \circ g^{-1} \circ h$, for any given $k > 0$ one has

$$h \circ f^{-k}(x^*) = g^{-k} \circ h(x^*)\,.$$

A comparison of the above two equalities shows that the orbit of x^* under iterates of f is mapped by h to the orbit of $h(x^*)$ under iterates of map g. This proves part (i).

The conclusion of part (ii) follows from a direct calculation:

$$\frac{df}{dx}\bigg|_{x=x^*} = \frac{dh^{-1}}{dx}\bigg|_{x=x^*} \cdot \frac{dg}{dx}\bigg|_{x=h(x^*)} \cdot \frac{dh}{dx}\bigg|_{x=x^*}\,,$$

noting that similar matrices have the same eigenvalues.   □

Example 1.14. The damped pendulum system (1.1), namely,

$$\dot{x} = y\,,$$
$$\dot{y} = -\frac{\kappa}{m}\,y - \frac{g}{\ell}\,\sin(x)\,,$$

has two fixed points:

$$(x^*, y^*) = (\theta^*, \dot{\theta}^*) = (0, 0) \quad \text{and} \quad (x^*, y^*) = (\theta^*, \dot{\theta}^*) = (\pi, 0)\,.$$

It is known from the pendulum physics (see Fig. 1.21) that the first fixed point is stable while the second, unstable.

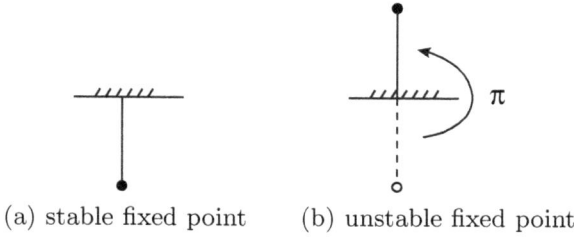

(a) stable fixed point   (b) unstable fixed point

Fig. 1.21   Two fixed points of the damped pendulum.

According to the above analysis, the Jacobian of the damped pendulum system is

$$J = \begin{bmatrix} 0 & 1 \\ -g\,\ell^{-1}\cos(x) & -\kappa/m \end{bmatrix}.$$

There are two cases to consider at the two fixed points:

(a) $x^* = \theta^* = 0$

In this case, the two eigenvalues of J are

$$\lambda_{1,2} = -\frac{\kappa}{2m} \pm \frac{1}{2}\sqrt{(\kappa/m)^2 - 4(g/\ell)}\,,$$

implying that the fixed point is stable since $\mathrm{Re}\{\lambda_{1,2}\} < 0$.

(b) $x^* = \theta^* = \pi$

In this case, the two eigenvalues of J are

$$\lambda_{1,2} = -\frac{\kappa}{2m} \pm \frac{1}{2}\sqrt{(\kappa/m)^2 + 4(g/\ell)}\,,$$

where $\mathrm{Re}\{\lambda_1\} > 0$ and $\mathrm{Re}\{\lambda_2\} < 0$, which implies that the fixed point is a saddle node and, hence, is unstable in one direction on the plane shown in Fig. 1.21, along which the pendulum can swing back and forth.

Clearly, the mathematical analysis given here is consistent with the physics of the damped pendulum as discussed before.

Example 1.15. For easy explanation of concept, consider a composite 2- and 1-dimensional autonomous system,

$$\dot{\mathbf{x}} = \begin{bmatrix} -1 & -2 & 0 \\ 2 & -1 & 0 \\ 0 & 0 & 1 \end{bmatrix} \mathbf{x}.$$

At the fixed point $(0, 0, 0)$, this system has eigenvalues $-1 \pm 2j$ and 1, implying that the origin is a saddle node, as illustrated by the phase portrait in Fig. 1.22.

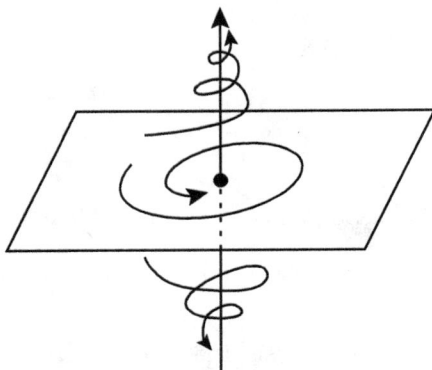

Fig. 1.22   3-dimensional saddle node.

Example 1.16. Consider a simplified coupled-neuron model,

$$\dot{x} = -\alpha x + h(\beta - y),$$
$$\dot{y} = -\alpha y + h(\beta - x),$$

where $\alpha > 0$ and $\beta > 0$ are constants, and $h(u)$ is a continuous function satisfying $h(-u) = -h(u)$ with $h'(u)$ being two-sided monotonically decreasing as $u \to \pm\infty$. One typical case is the sigmoidal function

$$h(u) = \frac{2}{1 - e^{-au}} - 1, \qquad a > 0.$$

In this coupled-neuron model, with the general function h as described, one has the following:

(i) there is a fixed point at $x^* = y^* := \lambda$;
(ii) if

$$h'(\beta - \lambda) = -\frac{dh(\beta - y)}{dy}\bigg|_{y=\lambda} < \alpha,$$

then this fixed point is unique and is a stable node;
(iii) if $h'(\beta - \lambda) > \alpha$, then there are two other fixed points, at

$$\left(\mu, \alpha^{-1} h(\beta - \mu)\right) \qquad \text{and} \qquad \left(\alpha^{-1} h(\beta - \mu), \mu\right)$$

respectively, for the same value of μ; they are stable nodes; but the one at (λ, λ) becomes a saddle point in this case.

Now it is noted that a fixed point at $x^* = y^* = \lambda$ is equivalent to showing that $-\alpha\lambda + h(\beta - \lambda) = 0$, or that the straight line $z = \alpha x$ and the curve $z = h(\beta - x)$ has a crossing point on the x–z plane. This is obvious from the geometry depicted in Fig. 1.23, since h is continuous.

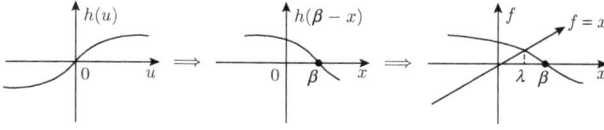

Fig. 1.23   Existence of a fixed point in the coupled-neuron model.

Then it is noted that λ is the unique root of equation

$$f(\lambda) := -\alpha\lambda + h(\beta - \lambda) = 0$$

being equivalent to showing that the function $f(\lambda)$ is strictly monotonic, so that $f(\lambda) = 0$ has only one root. To verify this, observe that

$$f'(\lambda) = -\alpha + h'(\beta - \lambda) < 0.$$

where $h'(\beta - \lambda) = -d\,h(\beta - \lambda)/d\lambda < \alpha$ by assumption. This implies that $f(\lambda)$ is decreasing. Moreover,

$$h(-u) = -h(u) \quad \Longrightarrow \quad -h'(-u) = -h'(u) \quad \Longrightarrow \quad h'(-u) = h'(u).$$

Since $h'(u)$ is two-sided monotonically decreasing as $u \to \pm\infty$, so are $h'(-u)$ and

$$f'(\lambda) = -\alpha + h'(\beta - \lambda).$$

Therefore, $f(\lambda)$ is strictly monotonic, so $f(\lambda) = 0$ has only one root.

To determine the stability of this root, by examining the Jacobian

$$J\Big|_{x=y=\lambda} = \begin{bmatrix} -\alpha & h'(\beta - \lambda) \\ h'(\beta - \lambda) & -\alpha \end{bmatrix},$$

one can see that its eigenvalues

$$s_1 = -\alpha - h'(\beta - \lambda) \quad \text{and} \quad s_2 = -\alpha + h'(\beta - \lambda)$$

satisfy $s_1 < s_2 < 0$, since $h'(\beta - \lambda) < \alpha$. Hence, $x = y = \lambda$ is a stable node.

Finally, consider the following equation:

$$f(x) = -\alpha x + h(\beta - x) = 0$$

on the x–z plane. If $h'(\beta - \lambda) > \alpha$, then it can be verified that the curve $z = h(\beta - x)$ and the straight line $z = \alpha x$ have three crossing points, as shown in Fig. 1.24.

In the above, it has already been shown that at least one crossing point is at $x = \lambda$, where $h'(\beta - \lambda) > \alpha$. It can be further verified that there must

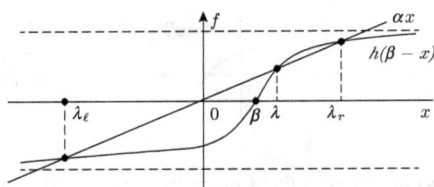

Fig. 1.24   Three crossing points between the curve and the straight line.

be two more crossing points, one at $\lambda_r > \lambda$ and the other at $\lambda_\ell < \lambda$, as depicted in Fig. 1.24.

Indeed, for $x > \lambda$, since $h'(\beta - x)$ is monotonically decreasing as discussed above, one has

$$f'(x) = -\alpha + h'(\beta - x) > -\alpha + h'(\beta - \lambda) > 0\,,$$

so that

$$h'(\beta - x) > \alpha > 0\,, \qquad \text{for all} \quad x > \alpha\,.$$

Since $h'(\beta - x)$ is two-sided monotonically decreasing, $h'(\beta - x) \to -\infty$ as $x \to \infty$, there must be a point, λ_r, such that $h'(\beta - \lambda_r) = \alpha$. This implies that the two curves $h(\beta - x)$ and αx have a crossing point $\lambda_r > \lambda$. The existence of another crossing point, $\lambda_\ell < \lambda$, can be similarly verified.

Next, to find the two new fixed points of the system, one can set

$$-\alpha x + h(\beta - y) = 0$$

to obtain

$$x_1^* = \alpha^{-1} h(\beta - \mu)\,,$$
$$y_1^* = \mu\,,$$

and set

$$-\alpha y + h(\beta - x) = 0$$

to obtain

$$x_2^* = \mu\,,$$
$$y_2^* = \alpha^{-1} h(\beta - \mu)\,,$$

where μ is a real value. These solutions have the same Jacobian, and the eigenvalues of the Jacobian are

$$s_1 = -\alpha - \sqrt{h'(\beta - x)h'(\beta - y)}\,,$$
$$s_2 = -\alpha + \sqrt{h'(\beta - x)h'(\beta - y)}\,,$$

which satisfy $s_1 < s_2 < 0$ at the above two crossing points, and satisfy $s_1 < 0 < s_2$ at $x^* = y^* = \lambda$, where the latter is a saddle node.

Now, return to the general nonlinear autonomous system (1.27).

Definition 1.6. The fixed point $\mathbf{x}^*$ of the autonomous system (1.27) is said to be *hyperbolic*, if all the eigenvalues of the system Jacobian J at this fixed point have nonzero real parts.

The importance of hyperbolic fixed points of a nonlinear autonomous system can be appreciated by the following fundamental result on the local dynamics of the autonomous system.

Theorem 1.6 (Grobman–Hartman Theorem for Systems). *Let* $\mathbf{x}^*$ *be a hyperbolic fixed point of the nonlinear autonomous system* (1.27). *Then, the dynamical properties of this system is qualitatively the same as that of its linearized system, in a (small) neighborhood of* $\mathbf{x}^*$.

Here, the equivalence of the dynamics of the two systems is local, and this theorem is not applicable to a nonautonomous system in general.

Proof. See [Robinson (1995)]: p. 158. □

For discrete-time systems, there is another version of the theorem for maps.

Theorem 1.7 (Grobman–Hartman Theorem for Maps). *Let* $\mathbf{x}^*$ *be a hyperbolic fixed point of the continuously differentiable map* $\mathbf{f} : R^n \to R^n$. *Then, the dynamical properties of this map is qualitatively the same as that of its linearized map* $[D\mathbf{f}(\mathbf{x}^*)] : R^n \to R^n$, *in a (small) neighborhood of* $\mathbf{x}^*$.

Proof. Without loss of generality, assume that $\mathbf{x}^* = 0$. Let $A = [D\mathbf{f}(0)]$, and decompose its state space according to the stable eigenvalues, denoted E^s, and unstable eigenvalues, denoted E^u, respectively. Then, $R^n = E^s \oplus E^u$. Denote $A_s = A\big|_{E^s}$ and $A_u = A\big|_{E^u}$, defined and restricted respectively on the two eigenspaces. By choosing appropriate coordinates, it can be assumed that $||A_s|| < 1$ and $||A_u^{-1}|| < 1 < ||A_u||$. Moreover, denote $\mu = \max\{||A_s||, ||A_u^{-1}||\} < 1$.

In a (small) neighborhood of the fixed point $\mathbf{x}^* = 0$, consider the expansion of the map $\mathbf{f}(\mathbf{x}) = [D\mathbf{f}(0)]\mathbf{x} + \mathbf{g}(\mathbf{x})$, with the higher-order terms satisfying $\mathbf{g}(0) = 0$ and $[D\mathbf{g}(0)] = 0$. Thus, for any small $\delta > 0$ there is a (small) neighborhood of the fixed point, in which $\sup_{\mathbf{x} \in R^n}\{||\mathbf{g}(\mathbf{x})|| + ||[D\mathbf{g}(\mathbf{x})]||\} < \delta$. This guarantees the existence of $\mathbf{f}^{-1}$, which is also continuously differentiable.

Now, the proof is carried out by verifying the topologically conjugate relationship shown in Fig. 1.20.

The objective is to find a homeomorphism $\mathbf{h} : R^n \to R^n$ in the form of $\mathbf{h} = I + \mathbf{k}$, where I is the identity map and $\mathbf{k} : R^n \to R^n$ is a continuous map, such that

$$\mathbf{h} \circ (A + \mathbf{g}) = A \circ \mathbf{h}.$$

It can be verified that the topologically conjugate relationship is equivalent to either

$$\mathbf{k} = -\mathbf{g} \circ (A + \mathbf{g})^{-1} + A \circ \mathbf{k} \circ (A + \mathbf{g})^{-1}$$

or

$$\mathbf{k} = A^{-1} \circ \mathbf{g} + A^{-1} \circ \mathbf{k} \circ (A + \mathbf{g}).$$

Now, in the small neighborhood of the fixed point $\mathbf{x}^* = 0$, define a map $F(\mathbf{k}, \mathbf{g}) = F_s(\mathbf{k}, \mathbf{g}) + F_u(\mathbf{k}, \mathbf{g})$, according to the above topologically conjugate relationship, as follows:

$$F_s(\mathbf{k}, \mathbf{g}) = -\mathbf{g}_s \circ (A + \mathbf{g})^{-1} + A_s \circ \mathbf{k}_s \circ (A + \mathbf{g})^{-1}$$

and

$$F_u(\mathbf{k}, \mathbf{g}) = A_u^{-1} \circ \mathbf{g}_u + A_u^{-1} \circ \mathbf{k}_u \circ (A + \mathbf{g}).$$

It can be verified that $F(\mathbf{k}, \mathbf{g})$ is continuous in $\mathbf{g}$, satisfying $F(0,0) = 0$, and

$$\|F(\mathbf{k}, \mathbf{g})\| \le \|\mathbf{g}\| + \mu\|\mathbf{k}\| \quad \text{and} \quad \|F(\mathbf{k}, \mathbf{g}) - F(\mathbf{k}', \mathbf{g})\| \le \mu\|\mathbf{k} - \mathbf{k}'\|,$$

for all continuous $\mathbf{k}, \mathbf{k}' \in R^n$. Therefore, $F(\mathbf{k}, \mathbf{g})$ is also continuous in $\mathbf{k}$ and furthermore $F(\cdot, \mathbf{g})$ is a uniform contraction mapping. Consequently, $F(\mathbf{k}', \mathbf{g}) = \mathbf{g}$ for some $\mathbf{k}'$ if and only if there is a $\mathbf{k}$ such that $\mathbf{k}' = \mathbf{k}(\mathbf{g})$. It follows that $\mathbf{h}(\mathbf{g}) = I + \mathbf{k}(\mathbf{g})$ and $\mathbf{h} \circ (A + \mathbf{g}) = A \circ \mathbf{h}$.

It remains to show that this $\mathbf{h} = \mathbf{h}(\mathbf{g})$ is a homeomorphism.

Consider $(A + \mathbf{g}) \circ \mathbf{r} = \mathbf{r} \circ A$ with $\mathbf{r} = I + \mathbf{k}'$. Similarly to the above, it can be verified that for each $\mathbf{g}$, correspondingly there exists a unique $\mathbf{r} = \mathbf{r}(\mathbf{g})$, perhaps for a smaller $\delta > 0$. It follows from the conjugate relationships for both $\mathbf{h}$ and $\mathbf{r}$ that

$$\begin{aligned}
\mathbf{h} \circ \mathbf{r} &= \left[A^{-1} \circ \mathbf{h} \circ (A + \mathbf{g})\right] \circ \mathbf{r} \\
&= A^{-1} \circ \mathbf{h} \circ \left[(A + \mathbf{g}) \circ \mathbf{r}\right] \\
&= A^{-1} \circ \mathbf{h} \circ \mathbf{r} \circ A.
\end{aligned}$$

On the other hand,

$$\mathbf{h} \circ \mathbf{r} = I + \mathbf{k}' + \mathbf{k} \circ [I + \mathbf{k}'] := I + \mathbf{s}.$$

Hence,

$$\mathbf{s} = A^{-1} \circ \mathbf{s} \circ A \qquad \text{if and only if} \qquad \mathbf{s} = A \circ \mathbf{s} \circ A^{-1}.$$

This implies that $F(\mathbf{s}, 0) = \mathbf{s}$. Since the fixed point is unique and since $F(0,0) = 0$, one has $\mathbf{s} = 0$; therefore, $\mathbf{h} \circ \mathbf{r} = I$. Similarly, it can be shown that $\mathbf{r} \circ \mathbf{h} = I$. Thus, $\mathbf{h}$ is a homeomorphism. □

The following example provides a visual illustration of Theorem 1.6.

Example 1.17. Consider a nonlinear system,

$$\dot{x} = -x \,,$$
$$\dot{y} = x^2 + y \,.$$

Its linearized system at $(0,0)$ is

$$\dot{x} = -x \,,$$
$$\dot{y} = y \,.$$

Their phase portraits are shown in Fig. 1.25.

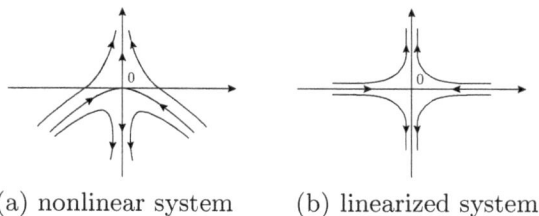

(a) nonlinear system   (b) linearized system

Fig. 1.25   Illustration of the Grobman–Hartman Theorem.

As can be seen, the two phase portraits are not exactly the same, but they are qualitatively the same, namely topologically equivalent, in the sense that one can be obtained from the other by smoothly bending the flow of the solution curves.

Exercises

1.1 For the following two linear systems, sketch by hand their phase portraits and classify their fixed points:

$$\dot{x} = -3x + 4y\,, \qquad \dot{y} = -2x + 3y\,,$$

and

$$\dot{x} = 4x - 3y\,, \qquad \dot{y} = 8x - 6y\,.$$

1.2 Consider the Duffing oscillator equation

$$\ddot{x}(t) + a\,\dot{x}(t) + b\,x(t) + c\,x^3(t) = \gamma\,\cos(\omega t)\,, \qquad (1.30)$$

where a, b, c are constants and $\gamma\,\cos(\omega t)$ is an external force input. By defining $y(t) = \dot{x}(t)$, rewrite this equation in a state-space form. Use a computer to plot its phase portraits for the following cases: $a = 0.4$, $b = -1.1$, $c = 1.0$, $\omega = 1.8$, and (1) $\gamma = 0.620$, (2) $\gamma = 1.498$, (3) $\gamma = 1.800$, (4) $\gamma = 2.100$, (5) $\gamma = 2.300$, (6) $\gamma = 7.000$. Indicate the directions of the orbit flows.

1.3 Consider the Chua circuit, shown in Fig. 1.26, which consists of one inductor (L), two capacitors (C_1, C_2), one linear resistor (R), and one piecewise-linear resistor (g).

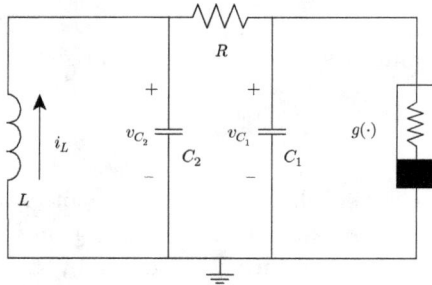

Fig. 1.26   The Chua circuit.

The dynamical equations of the circuit are

$$\begin{aligned}
C_1\,\dot{v}_{C_1} &= R^{-1}\big(v_{C_2} - v_{C_1}\big) - g(v_{C_1})\,, \\
C_2\,\dot{v}_{C_2} &= R^{-1}\big(v_{C_1} - v_{C_2}\big) + i_L\,, \\
L\,\dot{i}_L &= -v_{C_2}\,,
\end{aligned} \qquad (1.31)$$

where i_L is the current through the inductor L, v_{C_1} and v_{C_2} are the voltages across C_1 and C_2 respectively, and

$$g(v_{C_1}) = m_0 v_{C_1} + \frac{1}{2}(m_1 - m_0)\left(|v_{C_1} + 1| - |v_{C_1} - 1|\right),$$

with $m_0 < 0$ and $m_1 < 0$ being some appropriately chosen constants. This piecewise linear function is shown in Fig. 1.27.

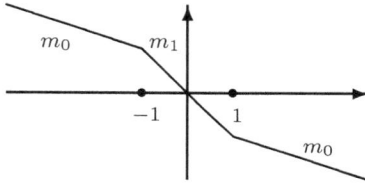

Fig. 1.27   The piecewise linear resistance in the Chua circuit.

Verify that, by defining $p = C_2/C_1 > 0$ and $q = C_2 R^2/L > 0$, with a change of variables, $x(\tau) = v_{C_1}(t)$, $y(\tau) = V_{C_2}(t)$, $z(\tau) = R i_L(t)$, and $\tau = t/(GC_2)$, the above circuit equations can be reformulated into the state-space form, as

$$\dot{x} = p\left(-x + y - f(x)\right),$$
$$\dot{y} = x - y + z,$$
$$\dot{z} = -qy, \tag{1.32}$$

where $f(x) = R g(v_{C_1})$.
For $p = 10.0$, $q = 14.87$, $m_0 = -0.68$, $m_1 = -1.27$, with initial conditions $(-0.1, -0.1, -0.1)$, use a computer to plot the circuit orbit portrait in the x–y–z space; or, show the portrait projections on the three principal planes: (a) the x–y plane, z–x the plane, and (c) the z–y plane.

1.4 Consider the following nonlinear system:

$$\dot{x} = y + \kappa x \left(x^2 + y^2\right),$$
$$\dot{y} = -x + \kappa y \left(x^2 + y^2\right).$$

Show that $(0, 0)$ is the only fixed point, and find under what condition on the constant κ, this fixed point is a stable or unstable focus. [Hint: Polar coordinates may be convenient to use.]

1.5 For the following two nonlinear systems, determine the types and the stabilities of their fixed points:

$$\ddot{y} + y + y^3 = 0\,,$$

and

$$\dot{x} = -x + xy\,,$$
$$\dot{y} = y - xy\,.$$

1.6 For each of the following systems, find the fixed points and determine their types and stabilities:

(a)

$$\dot{x} = y\cos(x)\,,$$
$$\dot{y} = \sin(x)\,;$$

(b)

$$\dot{x} = (x - y)\left(x^2 + y^2 - 1\right),$$
$$\dot{y} = (x + y)\left(x^2 + y^2 - 1\right);$$

(c)

$$\dot{x} = 1 - x\,y^{-1}\,,$$
$$\dot{y} = -x\,y^{-1}\left(1 - x\,y^{-1}\right);$$

(d)

$$\dot{x} = y\,,$$
$$\dot{y} = -x - \frac{1}{3}x^3 - y\,.$$

1.7 Let f and g be two topologically equivalent maps in metric spaces X and Y, and h be the homeomorphism satisfying $g = h^{-1} \circ f \circ h$. Verify that $h(g^k(x)) = f^k(h(x))$ for any integer $k \geq 0$.

1.8 Verify that the following two maps are not topologically equivalent in any neighborhood of the origin: $f(x) = x$ and $g(x) = x^2$.

Chapter 2

Stabilities of Nonlinear Systems (I)

Stability theory plays a central role in systems engineering, especially in the field of control systems and automation, with regard to both dynamics and control.

Stability of a dynamical system, with or without control and disturbance inputs, is a fundamental requirement in real-world applications. Roughly speaking, stability means that the system outputs and its internal signals are bounded within some allowable limits (the so-called bounded-input bounded-output stability) or, sometimes more strictly, the system outputs tend to an equilibrium state of interest (the so-called asymptotic stability). Conceptually, there are different kinds of stabilities, among which three basic notions are the main concerns in nonlinear dynamics and control systems: the stability of a system with respect to its equilibria, the orbital stability of the system output trajectory, and the structural stability of the system itself.

The basic concept of stability emerged from the study of an equilibrium state of a mechanical system, dated back to as early as 1644 when E. Torricelli studied the equilibrium of a rigid body under the natural force of gravity. The classical stability theorem of G. Lagrange, formulated in 1788, is perhaps the most well-known result about stability of conservative mechanical systems: for a conservative system currently at the position of an isolated equilibrium and perhaps subject to some simple constraints, if its potential energy has a minimum then this equilibrium position of the system is stable.

The evolution of the fundamental concepts of system stability and trajectory stability then went through a long history of development, with many advances and fruitful results, till the celebrated Ph.D. thesis of A. M. Lyapunov, "The General Problem of Motion Stability," completed in 1892. This monograph is so fundamental that its ideas and techniques are virtually leading all basic research and applications on stabilities of nonlinear dynamical systems today. In fact, not only dynamical behavioral analysis in modern physics but also controllers design in engineering systems depend upon the principles of Lyapunov's stability theories.

This chapter is devoted to a brief introduction to the basic stability theories, criteria, and methodologies of Lyapunov, as well as a few related important stability concepts, for general nonlinear dynamical systems (see [Chen (1999)] for a brief summary).

2.1 Lyapunov Stabilities

Briefly, the *Lyapunov stability* of a system with respect to its equilibrium of interest involves the behavior of the system outputs toward the equilibrium state, that are wandering nearby and around the equilibrium (*stability in the sense of Lyapunov*) or gradually approaching it (*asymptotic stability*); the orbital stability of a system output is the resistance of the trajectory to small perturbations; the structural stability of a system is the robustness of the system structural properties against small perturbations. These three basic types of stabilities are introduced in this section, for dynamical systems without explicitly involving control inputs.

To start with, consider a general nonautonomous system,

$$\dot{\mathbf{x}} = \mathbf{f}(\mathbf{x}, t), \qquad \mathbf{x}(t_0) = \mathbf{x}_0 \in R^n, \tag{2.1}$$

where, without loss of generality, it is assumed that the origin $\mathbf{x}^* = 0$ is a system fixed point. Lyapunov stability theory concerns various stabilities of the system trajectories with respect to this fixed point. When another fixed point is discussed, the new one is first shifted to zero by a change of variables, and then the transformed system is studied in the same way.

Definition 2.1. System (2.1) is said to be *stable in the sense of Lyapunov* about the fixed point $\mathbf{x}^* = 0$ if, for any $\varepsilon > 0$ and any initial time $t_0 \geq 0$, there exists a constant, $\delta = \delta(\varepsilon, t_0) > 0$, such that

$$||\mathbf{x}(t_0)|| < \delta \implies ||\mathbf{x}(t)|| < \varepsilon \qquad \text{for all } t \geq t_0. \tag{2.2}$$

This stability is said to be *uniform with respect to the initial time*, if the constant $\delta = \delta(\varepsilon)$ is independent of t_0 over the entire time interval $[0, \infty)$.

This concept of stability is illustrated in Fig. 2.1.

It should be emphasized that the constant δ generally depends on both ε and t_0. Only for autonomous systems is it always independent of t_0

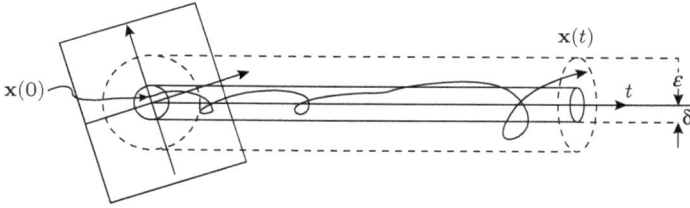

Fig. 2.1 Geometric meaning of stability in the sense of Lyapunov.

(thus, this stability for autonomous systems is always uniform with respect to the initial time). It is particularly important to point out that, unlike autonomous systems, one cannot simply fix the initial time $t_0 = 0$ for a nonautonomous system in a general discussion of its stability.

Example 2.1. Consider the following linear time-varying system with a discontinuous coefficient:

$$\dot{x}(t) = \frac{1}{1-t} x(t), \qquad x(t_0) = x_0.$$

It has an explicit solution

$$x(t) = x_0 \frac{1 - t_0}{1 - t}, \qquad 0 \le t_0 \le t < \infty,$$

which is well defined over the entire interval $[0, \infty)$ even when $t \to 1$ (see Fig. 2.2). Clearly, this solution is stable in the sense of Lyapunov about the equilibrium $x^* = 0$ over the entire time domain $[t_0, \infty)$ if and only if $t_0 \ge 1$.

This example shows that the initial time, t_0, indeed plays an important role in the stability of a nonautonomous system.

At this point, it is worth noting that it is often not desirable to shift the initial time t_0 to zero by a change of the time variable, $t \to t - t_0$ for control systems. For instance, consider a system that initially starts operation at time zero and then is subject to an external control input at time $t_0 > 0$. For this system, both controlled and uncontrolled behaviors are important in the analysis of the entire process of the system state evolution. Another example is the case where a system has multiple singularities or equilibria of interest, which are not allowed to change or cannot be eliminated by a shift using a single change of variables.

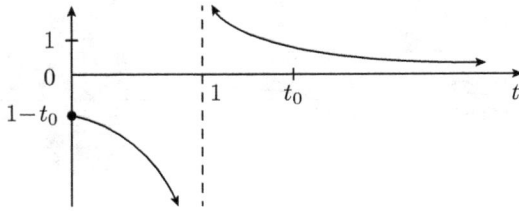

Fig. 2.2   Solution curve with a singularity at $t_0 = 1$.

It should also be noted that if a system is stable in the sense of Lyapunov about its fixed point, then starting from any bounded initial state in the neighborhood of the fixed point, its corresponding solution trajectories will always be bounded, which follows directly from the definition of the stability. Note that this is always true for linear systems, including time-varying systems. To show this, first recall that, for a linear time-varying system,

$$\dot{x}(t) = A(t)\,x(t)\,, \qquad x(t_0) = x_0 \in R^n\,, \tag{2.3}$$

the *fundamental matrix* is defined by

$$\Phi(t, t_0) = \exp\left\{ \int_{t_0}^{t} A(\tau)\,d\tau \right\}\,, \tag{2.4}$$

which satisfies $\Phi(s, s) = I$ for all $t_0 \le s \le t < \infty$. In particular, if $A(t) = A$ is a constant matrix then

$$\Phi(t, t_0) = e^{(t - t_0)A} := \Phi(t - t_0)\,.$$

Using matrix (2.4), any solution of system (2.3) can be expressed as

$$x(t) = \Phi(t, t_0)\,x_0\,. \tag{2.5}$$

Proposition 2.1. *System* (2.3) *is stable in the sense of Lyapunov about its zero fixed point if and only if all its solution trajectories are bounded.*

Proof. If the system is stable about its zero fixed point, then for any $\varepsilon > 0$, there exists a $\delta > 0$ such that $||\widetilde{x}_0|| < \delta$ implies that, starting from this $\widetilde{x}_0$ at time t_0, the system solution trajectory $x(t; t_0, \widetilde{x}_0)$ satisfies

$$\big|\big| \, x\big(t; t_0, \widetilde{x}_0\big) \, \big|\big| = \big|\big| \Phi(t, t_0)\widetilde{x}_0 \big|\big| < \varepsilon\,.$$

In particular, letting $\widetilde{x}_0 = [0 \ \cdots \ 0 \ \delta/2 \ 0 \ \cdots 0]^\top$, with $\delta/2$ being the ith component, yields

$$\big|\big| \Phi(t, t_0)\widetilde{x}_0 \big|\big| = \big|\big| \Phi_i(t, t_0) \big|\big|\, \delta/2 < \varepsilon\,,$$

where Φ_i is the ith column of Φ. Therefore, $||\Phi(t, t_0)|| < 2n\varepsilon\delta^{-1}$, so that

$$||\mathbf{x}(t; t_0, \mathbf{x}_0)|| = ||\Phi(t, t_0)\mathbf{x}_0|| < 2n\varepsilon\delta^{-1}||\mathbf{x}_0||,$$

which means that the solution trajectory $\mathbf{x}(t)$ is bounded if $\mathbf{x}_0$ is so.

Conversely, if all solution trajectories of system (2.3) are bounded, then there is a constant $c > 0$ such that $||\Phi(t, t_0)|| < c$. Thus, for any given $\varepsilon > 0$, $||\mathbf{x}_0|| < \delta := \varepsilon/c$ implies that

$$||\mathbf{x}(t; t_0)|| = ||\Phi(t, t_0)\mathbf{x}_0|| \leq c||\mathbf{x}_0|| < \varepsilon.$$

Hence, the system is stable in the sense of Lyapunov about its zero fixed point. $\qquad\square$

Next, a more important type of stability is introduced.

Definition 2.2. System (2.1) is said to be *asymptotically stable* about its fixed point $\mathbf{x}^* = 0$, if it is stable in the sense of Lyapunov and furthermore there exists a constant, $\delta = \delta(t_0) > 0$, such that

$$||\mathbf{x}(t_0)|| < \delta \implies ||\mathbf{x}(t)|| \to 0 \qquad \text{as} \quad t \to \infty. \tag{2.6}$$

The asymptotic stability is said to be *uniform*, if the constant δ is independent of t_0 over $[0, \infty)$, and is said to be *global* if the convergence $||\mathbf{x}|| \to 0$ is independent of the initial state $\mathbf{x}(t_0)$ over the entire spatial domain on which the system is defined (e.g. with $\delta = \infty$). If, furthermore,

$$||\mathbf{x}(t_0)|| < \delta \implies ||\mathbf{x}(t)|| \leq c e^{-\sigma t}, \quad t \geq t_0, \tag{2.7}$$

for two constants $c, \sigma > 0$, then the system is said to be *exponentially stable* about its fixed point $\mathbf{x}^*$.

The reason for the first requirement of being stable in the sense of Lyapunov is to exclude those unstable transient situations like the one shown in Example 2.1. The asymptotic stability is visualized in Fig. 2.3, and the exponential stability, in Fig. 2.4.

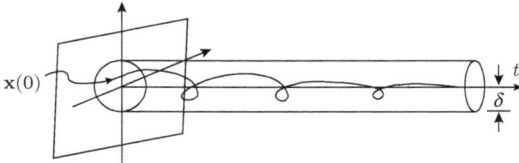

Fig. 2.3   Geometric meaning of the asymptotic stability.

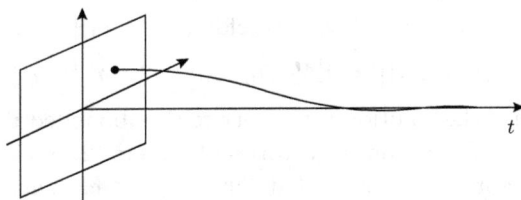

Fig. 2.4   Geometric meaning of the exponential stability.

Clearly, exponential stability implies asymptotic stability, and asymptotic stability implies the stability in the sense of Lyapunov, but the reverse is usually not true.

Example 2.2. For illustration, a system with output $x_1(t) = x_0 \sin(t)$ is stable in the sense of Lyapunov, but is not asymptotically stable, about 0; a system with output $x_2(t) = x_0 \left(1 + t - t_0\right)^{-1/2}$ is asymptotically stable (so, also is stable in the sense of Lyapunov), but is not exponentially stable about 0; however, a system with output $x_3(t) = x_0 \, e^{-t}$ is exponentially stable (hence, is both asymptotically stable and stable in the sense of Lyapunov). Finally, the simple example shown in Fig. 2.5 is asymptotically stable but not uniformly.

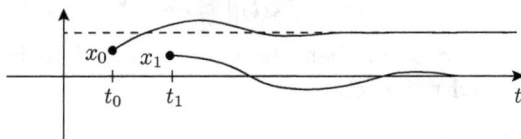

Fig. 2.5   Asymptotically but not uniformly stable.

2.2   Lyapunov Stability Theorems

Most stability theorems derived in this section apply to the general nonautonomous system (2.1), namely,

$$\dot{\mathbf{x}} = \mathbf{f}(\mathbf{x}, t), \qquad \mathbf{x}(t_0) = \mathbf{x}_0 \in R^n, \tag{2.8}$$

where $\mathbf{f} : \mathcal{D} \times [0, \infty) \to R^n$ is defined and continuously differentiable in a neighborhood of the origin, $\mathcal{D} \subseteq R^n$, with a given initial state $\mathbf{x}_0 \in \mathcal{D}$.

Again, without loss of generality, assume that $\mathbf{x}^* = 0$ is the system's fixed point of interest.

First, for the general autonomous system

$$\dot{\mathbf{x}} = \mathbf{f}(\mathbf{x}), \qquad \mathbf{x}(t_0) = \mathbf{x}_0 \in R^n, \tag{2.9}$$

an important special case of (2.8), with a continuously differentiable $\mathbf{f}$: $\mathcal{D} \to R^n$, the following criterion of stability, called the *first* (or *indirect*) *method of Lyapunov*, is very convenient to use.

Theorem 2.1 (First Method of Lyapunov).

[For continuous-time autonomous systems]

In system (2.9), *let* $J = \left[\partial \mathbf{f}/\partial \mathbf{x}\right]_{\mathbf{x}=\mathbf{x}^*=0}$ *be its Jacobian evaluated at the zero fixed point. If all the eigenvalues of J have negative real parts, then the system is asymptotically stable about* $\mathbf{x}^* = 0$.

Proof. This theorem is a direct consequence of the Grobman–Hartman Theorem, i.e. Theorem 1.6 introduced in Chapter 1, and the linear systems stability criteria. $\qquad\qquad\qquad\qquad\qquad\qquad\qquad\qquad\qquad\qquad\square$

Note that this and the following Lyapunov theorems apply to linear systems as well, for linear systems are merely special cases of nonlinear systems. When $\mathbf{f}(\mathbf{x}) = A\mathbf{x}$, the linear time-invariant system $\dot{\mathbf{x}} = A\mathbf{x}$ has the only fixed point $\mathbf{x}^* = 0$, and the system Jacobian is $J = A$. Thus, if A has all eigenvalues with negative real parts, the system is asymptotically stable about its fixed point. Therefore, Theorem 2.1 is consistent with the familiar linear stability theory.

Note also that the region of asymptotic stability given by Theorem 2.1 is local, which can be quite large for some nonlinear systems but may be very small for some others. However, there is no general criterion for determining the boundary of such a local stability region, which is also called the *region of attraction*, when this and the following Lyapunov methods are applied. Clearly, when it is applied to linear systems, the stability is always global.

Next, before introducing the next theorem of Lyapunov, two examples are first discussed.

Example 2.3. Consider the damped pendulum (1.1), namely,

$$\dot{x} = y$$
$$\dot{y} = -\frac{g}{\ell}\sin(x) - \frac{\kappa}{m}y.$$

Its Jacobian is

$$J = \begin{bmatrix} 0 & 1 \\ -g\,\ell^{-1}\cos(x) & -\kappa/m \end{bmatrix}_{(x^*,y^*)=(0,0)} = \begin{bmatrix} 0 & 1 \\ -g\,\ell^{-1} & -\kappa/m \end{bmatrix},$$

which has eigenvalues

$$\lambda_{1,2} = -\frac{\kappa}{2m} \pm \frac{1}{2}\sqrt{(\kappa^2/m^2) - 4g/\ell},$$

both with negative real parts. Hence, the damped pendulum system is asymptotically stable about its fixed point $(x^*, y^*) = (\theta^*, \dot{\theta}^*) = (0,0)$.

It is very important to note that Theorem 2.1 cannot be applied to a general nonautonomous system, because this theorem is neither necessary nor sufficient, as shown by the following example.

Example 2.4. Consider the following linear time-varying system:

$$\dot{\mathbf{x}}(t) = \begin{bmatrix} -1 + 1.5\cos^2(t) & 1 - 1.5\sin(t)\cos(t) \\ -1 - 1.5\sin(t)\cos(t) & -1 + 1.5\sin^2(t) \end{bmatrix} \mathbf{x}(t).$$

This system has eigenvalues $\lambda_{1,2} = -0.25 \pm j\,0.25\sqrt{7}$, both with negative real parts and being independent of the time variable t. If Theorem 2.1 is used to judge this system, the conclusion would be that the system is asymptotically stable about its fixed point 0.

However, the solution of this system is precisely

$$\mathbf{x}(t) = \begin{bmatrix} e^{0.5t}\cos(t) & e^{-t}\sin(t) \\ -e^{0.5t}\sin(t) & e^{-t}\cos(t) \end{bmatrix} \begin{bmatrix} x_1(t_0) \\ x_2(t_0) \end{bmatrix},$$

which is unstable for any initial conditions with a bounded and nonzero value of $x_1(t_0)$, no matter how small this initial value is.

This example shows that by using the Lyapunov first method alone to determine the stability of a general time-varying system, the conclusion can be wrong.

A rigorous approach for asymptotic stability analysis of general nonautonomous systems is provided by the second method of Lyapunov. The concept of class-$\mathcal{K}$ function will be useful, where $\mathcal{K}$ refers to its first use of the Greek letter κ, which is defined as follows:

$$\mathcal{K} = \{\, g(t): \quad g(t_0) = 0, \quad g(t) > 0 \ \text{if } t > t_0,$$
$$g(t) \text{ is continuous and nondecreasing on } [t_0, \infty) \,\}.$$

Theorem 2.2 (Second Method of Lyapunov).

[For continuous-time nonautonomous systems]

System (2.8) is globally (over the entire domain $\mathcal{D}$), uniformly (with respect to the initial time over the entire time interval $[t_0, \infty)$), and asymptotically stable about its zero fixed point, if there exist a scalar-valued function, $V(\mathbf{x}, t)$, defined on $\mathcal{D} \times [t_0, \infty)$, and three functions $\alpha(\cdot), \beta(\cdot), \gamma(\cdot) \in \mathcal{K}$, such that

(i) $V(0, t) = 0$ *for all* $t \geq t_0$;
(ii) $V(\mathbf{x}, t) > 0$ *for all* $\mathbf{x} \neq 0$ *in* $\mathcal{D}$ *and all* $t \geq t_0$;
(iii) $\alpha(||\mathbf{x}||) \leq V(\mathbf{x}, t) \leq \beta(||\mathbf{x}||)$ *for all* $t \geq t_0$;
(iv) $\dot{V}(\mathbf{x}, t) \leq -\gamma(||\mathbf{x}||) < 0$ *for all* $t \geq t_0$, $||\mathbf{x}|| \neq 0$.

In Theorem 2.2, the function V is called a *Lyapunov function*, and the method of constructing a Lyapunov function for stability determination is called the *second* (or *direct*) *method of Lyapunov*.

The role of Lyapunov function in the theorem is illustrated in Fig. 2.6, where for simplicity only the autonomous case is visualized. In the figure, assume that a Lyapunov function, $V(\mathbf{x})$, has been found, with a bowl-shape as shown based on conditions (i) and (ii). Then, condition (iv) is

$$\dot{V}(\mathbf{x}) = \left[\frac{\partial V}{\partial \mathbf{x}}\right]\dot{\mathbf{x}} < 0, \quad \mathbf{x} \neq 0, \tag{2.10}$$

where $\left[\frac{\partial V}{\partial \mathbf{x}}\right]$ is the gradient of V along the trajectory $\mathbf{x}$.

The operation of computing $\dot{V}(\mathbf{x})$ is usually said to take the derivative of $V(\mathbf{x})$ along the system trajectory $\mathbf{x}$.

Now, as is well known from Calculus, if the inner product of the above gradient and the tangent vector $\dot{\mathbf{x}}$ is constantly negative, as guaranteed by condition (2.10), then the angle between these two vectors is larger than $\pi/2$. This means that the surface of $V(\mathbf{x})$ is monotonically shrinking to zero (as seen in Fig. 2.6). Consequently, the system trajectory $\mathbf{x}$, with the projection on the domain as shown in the figure, converges to zero as time tends to infinity.

Proof. First, the stability in the sense of Lyapunov is established.

This requires to show that, for any $\varepsilon > 0$, there is a $\delta = \delta(\varepsilon, t_0) > 0$, such that

$$||\mathbf{x}(t_0)|| < \delta \quad \Longrightarrow \quad ||\mathbf{x}(t)|| < \varepsilon \quad \text{for all } t \geq t_0.$$

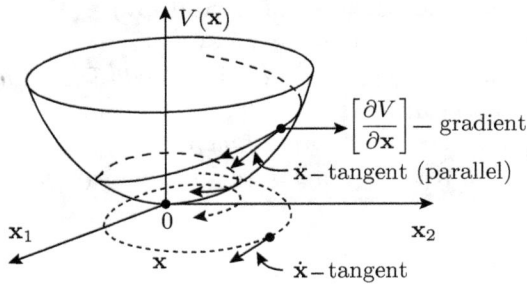

Fig. 2.6   Geometric meaning of the Lyapunov function.

The given conditions (iii) and (iv) together imply that, for $\mathbf{x} \neq 0$,

$$0 < \alpha(||\mathbf{x}||) \leq V(\mathbf{x}, t) \leq V\big(\mathbf{x}(t_0), t_0\big) \quad \text{for all } t \geq t_0 \,.$$

Since $V(\mathbf{x}, t)$ is continuous and satisfies $V(0, t_0) = 0$, there exists a $\delta = \delta(\varepsilon, t_0) > 0$ such that

$$||\mathbf{x}(t_0)|| < \delta \quad \Longrightarrow \quad V\big(\mathbf{x}(t_0), t_0\big) < \alpha(\varepsilon) \,,$$

because $\alpha(||\mathbf{x}||) > 0$ for $\mathbf{x} \neq 0$. Therefore, if $||\mathbf{x}(t_0)|| < \delta$, then

$$\alpha(||\mathbf{x}||) \leq V\big(\mathbf{x}(t_0), t_0\big) \leq \alpha(\varepsilon) \quad \text{for all } t \geq t_0 \,.$$

Since $\alpha(\cdot) \in \mathcal{K}$, this implies that $||\mathbf{x}(t)|| < \varepsilon$ for all $t \geq t_0$.

Next, the uniform stability property is verified.

The given condition (iii), namely,

$$0 < \alpha\,(||\mathbf{x}||) \leq V(\mathbf{x}, t) \leq \beta\,(||\mathbf{x}||) \quad \text{for all } \mathbf{x} \neq 0 \text{ and } t \geq t_0 \,,$$

implies that, for any $\varepsilon > 0$, there is a $\delta = \delta(\varepsilon) > 0$, independent of t_0, such that $\beta(\delta) < \alpha(\varepsilon)$, as illustrated in Fig. 2.7. Therefore, for any initial state $\mathbf{x}(t_0)$ satisfying $||\mathbf{x}(t_0)|| < \delta$,

$$\alpha(||\mathbf{x}(t)||) \leq V(\mathbf{x}, t) \leq V\big(\mathbf{x}(t_0), t_0\big) \leq \beta\big(||\mathbf{x}(t_0)||\big) \leq \beta(\delta) < \alpha(\varepsilon)$$

for all $t \geq t_0$, which implies that

$$||\mathbf{x}(t)|| < \varepsilon \quad \text{for all } t \geq t_0 \,.$$

Since $\delta = \delta(\varepsilon)$ is independent of t_0, this stability is uniform with respect to the initial time.

Further, the uniform asymptotic stability is verified.

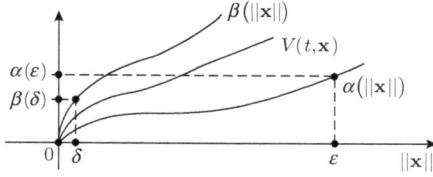

Fig. 2.7   Geometric meaning of the comparison $\beta(\delta) < \alpha(\varepsilon)$.

It follows from the uniform stability, just shown above, that for any $\varepsilon > 0$, there is a $\delta = \delta(\varepsilon) > 0$, independent of t_0, such that

$$||\mathbf{x}(t_0)|| < \delta \quad \Longrightarrow \quad ||\mathbf{x}(t)|| < \varepsilon \quad \text{for all} \ \ t \geq t_0 \,.$$

It is required to show that there is a $t^* > 0$, independent of t_0, such that

$$||\mathbf{x}(t_0)|| < \delta \quad \Longrightarrow \quad ||\mathbf{x}(t)|| < \varepsilon \quad \text{for all} \ \ t \geq t^* \,.$$

To do so, let

$$t^* = t^*(\varepsilon, \delta) = \beta(\delta)/\gamma(\delta) \,.$$

Then, it can be verified that there is a $t_1 \in (t_0, t_0 + t^*]$ such that

$$||\mathbf{x}(t_0)|| < \delta \quad \Longrightarrow \quad ||\mathbf{x}(t_1)|| < \delta \,.$$

If not, namely, $||\mathbf{x}(t_0)|| < \delta$ but $||\mathbf{x}(t)|| \geq \delta$ for all $t \in (t_0, t_0 + t^*]$, then condition (iv) implies that

$$\dot{V}(\mathbf{x}, t) \leq -\gamma(||\mathbf{x}||) \leq -\gamma(\delta) < 0 \,,$$

so that, by condition (iii),

$$V(\mathbf{x}(t), t) = V(\mathbf{x}(t_0), t_0) + \int_{t_0}^{t} \dot{V}(\mathbf{x}(\tau), \tau) \, d\tau$$

$$\leq V(\mathbf{x}(t_0), t_0) - \int_{t_0}^{t} \gamma(\delta) \, d\tau$$

$$= V(\mathbf{x}(t_0), t_0) - \gamma(\delta) \, (t - t_0)$$

$$\leq \beta(\delta) - \gamma(\delta) \, (t - t_0) \,.$$

Thus, it follows that, for all $\mathbf{x}(t)$ satisfying $||\mathbf{x}(t)|| \geq \delta$,

$$V(\mathbf{x}, t) \big|_{t=t_0+t^*} \leq \beta(\delta) - \gamma(\delta) \, t^* = \beta(\delta) - \beta(\delta) = 0 \,.$$

This contradicts the fact that $V(\mathbf{x}, t) > 0$ for all $\mathbf{x} \neq 0$. Therefore, as claimed above, there is a $t_1 \in (t_0, t_0 + t^*]$ such that, $||\mathbf{x}(t_0)|| < \delta$ implies $||\mathbf{x}(t_1)|| < \delta$. Consequently,

$$||\mathbf{x}(t_0)|| < \delta \quad \Longrightarrow \quad ||\mathbf{x}(t)|| < \varepsilon \quad \text{for all} \ \ t > t_0 + t^* \geq t_1 \,.$$

Since $t^* = t^*(\varepsilon, \delta)$ is independent of t_0, this asymptotic stability is uniform with respect to the initial time.

Finally, the global uniform asymptotic stability is proved.

Since $\alpha(||\mathbf{x}||) \to \infty$ as $||\mathbf{x}|| \to \infty$, one has

$$\alpha(||\mathbf{x}||) \le V(\mathbf{x}(t), t) \le \beta(||\mathbf{x}||) \to \infty \qquad \text{as} \quad ||\mathbf{x}|| \to \infty.$$

Hence, starting from any initial state $\mathbf{x}(t_0) \in R^n$, there is always a (large enough) $\delta > 0$ such that $||\mathbf{x}(t_0)|| < \delta$. For this δ, since $\alpha(||\mathbf{x}||) \to \infty$ as $||\mathbf{x}|| \to \infty$, there is always a (large enough) $r > 0$ such that $\beta(\delta) < \alpha(r)$ (see also Fig. 2.7). Thus,

$$\alpha(||\mathbf{x}(t)||) \le V(\mathbf{x}(t), t) \le V(\mathbf{x}(t_0), t_0) \le \beta(||\mathbf{x}(t_0)||) \le \beta(\delta) \le \alpha(r),$$

which implies that

$$||\mathbf{x}(t)|| < r \quad \text{for all } t \ge t_0.$$

This establishes the global stability in the sense of Lyapunov.

Under all the given conditions, the uniform asymptotic stability can also be verified in the same manner by repeating the above arguments. $\qquad \square$

Regarding the uniform negative definiteness condition (iv) of Theorem 2.2, it is very important to note that it cannot be weakened to be the simpler condition that $\dot{V}(\mathbf{x}, t) < 0$ for all $t \ge t_0$. The reason is that this weaker condition is not sufficient for nonautonomous systems in general, as can be seen from the following example.

Example 2.5. Consider the simplest 1-dimensional linear system,

$$\dot{x} = 0, \qquad t \ge 0.$$

Suppose that one chooses the Lyapunov function

$$V(x, t) = \frac{t+2}{t+1} x^2,$$

which satisfies

$$\dot{V}(x, t) = -\frac{x^2}{(t+1)^2} < 0 \qquad \text{for all } t \ge 0.$$

Then, one tends to conclude that the given system is asymptotically stable about its zero fixed point. However, the solution of this system is $x = x_0$, the constant initial condition, which is stable in the sense of Lyapunov but not asymptotically.

Now, return to Theorem 2.2. Since autonomous systems are the special case of nonautonomous systems, this theorem applies to them as well. The following three examples show how the theorem can be applied to the technical stability analysis of both autonomous and nonautonomous systems.

Example 2.6. Consider a general linear time-varying system,

$$\dot{\mathbf{x}}(t) = A(t)\,\mathbf{x}(t)\,,$$

where the system matrix $A(t)$ is assumed to satisfy the following condition: there are constants a and b and a uniformly positive definite and symmetric matrix $Q(t) \geq \lambda_{\min}(Q)I > 0$ for all $t \geq t_0$, where $\lambda_{\min}(Q)$ is the minimum eigenvalue of $Q(t)$, such that the time-varying Lyapunov equation

$$\dot{P}(t) + P(t)A(t) + A^\top(t)P(t) + Q(t) = 0$$

has a positive definite matrix solution $P(t)$ satisfying

$$0 < aI \leq P(t) \leq bI \qquad \text{for all } t \geq t_0\,.$$

It can be shown that this system is globally, uniformly, and asymptotically stable about its zero fixed point. Indeed, using the Lyapunov function

$$V(\mathbf{x}, t) = \mathbf{x}^\top(t)P(t)\,\mathbf{x}(t)\,,$$

and choosing the following three class-$\mathcal{K}$ functions:

$$\alpha(\tau) = a\,\tau^2\,, \qquad \beta(\tau) = b\,\tau^2\,, \qquad \gamma(\tau) = \lambda_{\min}(Q)\,\tau^2\,,$$

one has

$$\alpha(||\mathbf{x}||) = a\,||\mathbf{x}||^2 \leq V(\mathbf{x}, t) \leq b\,||\mathbf{x}||^2 = \beta(||\mathbf{x}||)\,,$$

and

$$\begin{aligned}
\dot{V}(\mathbf{x}, t) &= \mathbf{x}^\top \dot{P}\mathbf{x} + \mathbf{x}^\top P \dot{\mathbf{x}} + \dot{\mathbf{x}}^\top P \mathbf{x} \\
&= \mathbf{x}^\top \big[\dot{P} + PA + A^\top P\big]\mathbf{x} \\
&= -\mathbf{x}^\top Q\mathbf{x} \\
&\leq -\lambda_{\min}(Q)\,||\mathbf{x}||^2 \\
&= -\gamma(||\mathbf{x}||)\,.
\end{aligned}$$

Moreover,

$$\alpha(||\mathbf{x}||) \to \infty \qquad \text{as } ||\mathbf{x}|| \to \infty\,.$$

Therefore, all conditions stated in Theorem 2.2 are satisfied, so the system is globally, uniformly, and asymptotically stable about its zero fixed point.

Since a linear time-invariant system,

$$\dot{\mathbf{x}} = A\mathbf{x}, \qquad \mathbf{x}_0 \in R^n,$$

is a special case of the above linear time-varying system, this time-invariant system is asymptotically stable about its zero fixed point if and only if, for any positive definite matrix Q, the Lyapunov equation

$$A^\top P + PA + Q = 0$$

has a unique positive definite matrix solution P. In fact, this amounts to noting that for this constant matrix P, one has $\dot{P} = 0$ in the Lyapunov equation of the above example.

Example 2.7. The damped pendulum (1.1), namely,

$$\dot{x} = y$$
$$\dot{y} = -\frac{g}{\ell}\sin(x) - \frac{\kappa}{m}y$$

is asymptotically stable about its fixed point $(x^*, y^*) = (\theta^*, \dot{\theta}^*) = (0, 0)$. This can be verified by using the Lyapunov function

$$V(x, y) = \frac{1}{2}\begin{bmatrix} x & y \end{bmatrix} Q \begin{bmatrix} x \\ y \end{bmatrix} + \frac{g}{\ell}\left(1 - \cos(x)\right),$$

with

$$Q := \begin{bmatrix} \kappa^2/(2m^2) & \kappa/(2m) \\ \kappa/(2m) & 1 \end{bmatrix}.$$

Indeed, $V(x, y) > 0$ for all $-\pi/2 < x < \pi/2$ and all $y \in R$. Moreover,

$$\dot{V}(x, y) = \begin{bmatrix} x & y \end{bmatrix} Q \begin{bmatrix} \dot{x} \\ \dot{y} \end{bmatrix} + \frac{g}{\ell}\sin(x)\,\dot{x}$$

$$= -\frac{g\,\kappa}{2m\,\ell}x\sin(x) - \frac{\kappa}{m}y^2 < 0$$

for all $-\pi/2 < x < \pi/2$ and $y \in R$. By Theorem 2.2, this damped pendulum system is asymptotically stable about its fixed point $(0, 0)$.

Note that, although this damped pendulum is an autonomous system, Theorem 2.2 can still be applied, with

$$\alpha(\cdot) = \frac{1}{2}\lambda_{\min}(Q)\,(\cdot)^2,$$

$$\beta(\cdot) = \left(\frac{1}{2}\lambda_{\max}(Q) + 2\frac{g}{L}\right)(\cdot)^2,$$

$$\gamma(\mathbf{x}) = \min\left\{\frac{g}{2m\ell}, \frac{1}{m}\right\}\kappa\left(\sin(x) + y^2\right),$$

where $\kappa\sin(x)$ is a monotone function on $[0, \pi/2)$.

Before discussing one more example, a simple and useful result is presented.

Lemma 2.1. *For an autonomous system* $\dot{\mathbf{x}} = \mathbf{f}(\mathbf{x})$, *all the eigenvalues of its Jacobian,* J, *have negative real parts if and only if the Lyapunov equation*

$$J^\top P + PJ + Q = 0$$

has a unique positive definite matrix solution P *for any positive definite matrix* Q. *Moreover, a Lyapunov function for analyzing the system stability can be chosen as*

$$V(\mathbf{x}) = \mathbf{x}^\top P \mathbf{x}.$$

Proof. The first part of the lemma is well known in the linear systems theory.

If such a positive definite matrix P indeed exists, then $V(\mathbf{x})$ so constructed is a Lyapunov function, which satisfies $V(0) = 0$, $V(\mathbf{x}) > 0$ for all $\mathbf{x} \neq 0$, and

$$
\begin{aligned}
\dot{V}(\mathbf{x}) &= \mathbf{x}^\top P \dot{\mathbf{x}} + \dot{\mathbf{x}} P \mathbf{x} \\
&= \mathbf{x}^\top P \mathbf{f}(\mathbf{x}) + \mathbf{f}(\mathbf{x}) P \mathbf{x} \\
&= \mathbf{x}^\top P \left[J\mathbf{x} + \text{HOT} \right] + \left[J\mathbf{x} + \text{HOT} \right]^\top P \mathbf{x} \\
&= \mathbf{x}^\top \left[PJ + JP \right] \mathbf{x} + 2\mathbf{x}^\top P \left[\text{HOT} \right] \\
&= -\mathbf{x}^\top Q \mathbf{x} + 2\mathbf{x}^\top P \left[\text{HOT} \right] \\
&< 0 \qquad \text{for small enough } ||\mathbf{x}||,
\end{aligned}
$$

where J is the Jacobian and HOT represents the higher-order terms in the Taylor expansion of $\mathbf{f}(\mathbf{x})$ about the zero fixed point. □

Example 2.8. Consider a nonautonomous system in the special form of

$$\dot{\mathbf{x}} = A\mathbf{x} + \mathbf{g}(\mathbf{x}, t),$$

where A is a stable constant matrix and $\mathbf{g}$ is a nonlinear function satisfying $\mathbf{g}(0, t) = 0$ and $||\mathbf{g}(\mathbf{x}, t)|| \leq c\,||\mathbf{x}||$ for a constant $c > 0$ and for all $t \in [t_0, \infty)$.

Since A is stable, the following Lyapunov equation

$$PA + A^\top P + I = 0$$

has a unique positive definite matrix solution, P. Then, the Lyapunov function $V(\mathbf{x}, t) = \mathbf{x}^\top P \mathbf{x}$ yields

$$
\begin{aligned}
\dot{V}(\mathbf{x}, t) &= \mathbf{x}^\top P \dot{\mathbf{x}} + \dot{\mathbf{x}}^\top P \mathbf{x} \\
&= \mathbf{x}^\top \left[PA + A^\top P \right] \mathbf{x} + 2\mathbf{x}^\top P \mathbf{g}(\mathbf{x}, t) \\
&\leq -\mathbf{x}^\top \mathbf{x} + 2\lambda_{\max}(P)\, c\,||\mathbf{x}||^2,
\end{aligned}
$$

where $\lambda_{\max}(P)$ is the largest eigenvalue of P. Therefore, if the constant $c < 1/(2\lambda_{\max}(P))$ and if the following three class-$\mathcal{K}$ functions are used:

$$\alpha(\zeta) = \lambda_{\min}(P)\,\zeta^2, \quad \beta(\zeta) = \lambda_{\max}(P)\,\zeta^2,$$

$$\gamma(\zeta) = \left[1 - 2c\,\lambda_{\max}(P)\right]\zeta^2,$$

then conditions (iii) and (iv) of Theorem 2.2 are satisfied. As a result, the given system is globally, uniformly, and asymptotically stable about its zero fixed point.

This example shows that the linear part of a weakly nonlinear nonautonomous system can dominate the system stability.

Example 2.8 motivates the following investigation. Consider a nominal nonautonomous system,

$$\dot{\mathbf{x}} = \mathbf{f}(\mathbf{x}, t), \qquad \mathbf{x}(t_0) = \mathbf{x}_0 \in R^n,$$

where $\mathbf{f}$ is continuously differentiable, with a zero fixed point $\mathbf{f}(0, t) = 0$ for all $t \geq t_0$. Suppose that a small perturbation, $\mathbf{e}(\mathbf{x}, t)$, is input, yielding a perturbed system of the form

$$\dot{\mathbf{x}} = \mathbf{f}(\mathbf{x}, t) + \mathbf{e}(\mathbf{x}, t).$$

The question is: if the nominal system was asymptotically stable about its zero fixed point, then under a small perturbation does the perturbed system remain to be stable, at least in the sense of Lyapunov, about the zero fixed point? If so, the nominal system is said to be *totally stable* about its zero fixed point. This issue will be further addressed in the next chapter, for which the Malkin Theorem will provide an answer: if the nominal system is uniformly asymptotically stable about its zero fixed point, and the perturbation is small enough, then the perturbed system is totally stable.

At this point, it should be noted that in Theorem 2.2, the uniform stability is guaranteed by the class-$\mathcal{K}$ functions α, β, and γ stated in conditions (iii) and (iv), which is necessary since the solution of a nonautonomous system may sensitively depend on the initial time, as seen from Example 2.1. For autonomous systems, however, these class-$\mathcal{K}$ functions (hence, condition (iii)) are not needed. In this case, Theorem 2.2 reduces to the following simple form.

Theorem 2.3 (Second Method of Lyapunov).

[For continuous-time autonomous systems]

The autonomous system (2.9) *is globally (over the entire domain $\mathcal{D}$) and asymptotically stable about its zero fixed point, if there exists a scalar-valued function, $V(\mathbf{x})$, defined on $\mathcal{D}$, such that*

Table 2.1   Lyapunov functions for autonomous and nonautonomous systems.

Autonomous Systems	Nonautonomous Systems
$V(0) = 0$	$V(0, t) = 0$ for all $t \geq t_0$
$V(\mathbf{x}) > 0$	$V(\mathbf{x}, t) > 0$ for all $\mathbf{x} \neq 0$ and all $t \geq t_0$
	$0 < \alpha(||\mathbf{x}||) \leq V(\mathbf{x}, t) \leq \beta(||\mathbf{x}||)$ for some $\alpha, \beta \in \mathcal{K}$ and all $t \geq t_0$
$\dot{V}(\mathbf{x}) < 0$	$\dot{V}(\mathbf{x}, t) \leq -\gamma(||\mathbf{x}||) < 0$ for some $\gamma \in \mathcal{K}$ and all $t \geq t_0$

(i) $V(0) = 0;$

(ii) $V(\mathbf{x}) > 0$ *for all* $\mathbf{x} \neq 0$ *in* $\mathcal{D};$

(iii)   (not needed)

(iv) $\dot{V}(\mathbf{x}) < 0$ *for all* $\mathbf{x} \neq 0$ *in* $\mathcal{D}.$

Note that if condition (iv) in Theorem 2.3 is replaced by

(iv') $\dot{V}(\mathbf{x}) \leq 0$ for all $\mathbf{x} \in \mathcal{D}$,

then the resulting stability is only in the sense of Lyapunov but may not be asymptotic.

Note also that if $\mathcal{D} = R^n$, namely if $V(\mathbf{x}) \to \infty$ as $||\mathbf{x}|| \to \infty$, then the stability is global over R^n.

A comparison of the stability conditions for autonomous and nonautonomous systems stated in Theorems 2.2 and 2.3 is given in Table 2.1.

Example 2.9. Consider a linear time-invariant system,

$$\dot{\mathbf{x}} = A\mathbf{x},$$

with a negative definite constant matrix A.

If Theorem 2.1 is applied, one can use the Lyapunov function $V(x) = \frac{1}{2}\mathbf{x}^\top \mathbf{x}$, which satisfies $V(0) = 0$, $V(\mathbf{x}) > 0$ for all $\mathbf{x} \neq 0$, and

$$\dot{V}(\mathbf{x}) = \mathbf{x}\dot{\mathbf{x}} = \mathbf{x}^\top A\mathbf{x} < 0, \qquad \text{for all} \quad \mathbf{x} \neq 0.$$

Therefore, the system is asymptotically stable about its zero fixed point, consistent with the well-known linear systems theory.

Example 2.10. Consider the undamped pendulum (1.2), namely,

$$\dot{x} = y$$
$$\dot{y} = -\frac{g}{\ell}\sin(x),$$

where $x = \theta$ is the angular variable defined on $-\pi < \theta < \pi$, with the vertical axis as its reference, and g is the gravity constant.

It is easy to verify that the system Jacobian at the zero equilibrium has a pair of purely imaginary eigenvalues, $\lambda_{1,2} = \pm j \sqrt{g/\ell}$, where $j = \sqrt{-1}$ so that Theorem 2.1 is not applicable. However, if one uses the Lyapunov function

$$V = \frac{g}{\ell} \left(1 - \cos(x) \right) + \frac{1}{2} y^2 \,,$$

then

$$\begin{aligned}
\dot{V}(x) &= \frac{g}{\ell} \, \dot{x} \, \sin(x) + y \, \dot{y} \\
&= \frac{g}{\ell} \, y \, \sin(x) - \frac{g}{\ell} \, y \, \sin(x) = 0
\end{aligned}$$

for all (x, y) over the entire domain.

Thus, the conclusion is that this undamped pendulum is stable in the sense of Lyapunov but not asymptotically, consistent with the physics of the undamped pendulum.

Theorem 2.4 (Krasovskii Theorem).

[For continuous-time autonomous systems]

For the autonomous system (2.9), let $J(\mathbf{x}) = \left[\partial \mathbf{f} / \partial \mathbf{x} \right]$ be its Jacobian, which generally is a function of $\mathbf{x}(t)$. A sufficient condition for the system to be asymptotically stable about its zero fixed point is that there exist two real positive definite constant matrices, P and Q, such that the matrix

$$J^\top(\mathbf{x}) \, P + P \, J(\mathbf{x}) + Q$$

is semi-negative definite for all $\mathbf{x} \neq 0$ in a neighborhood $\mathcal{D}$ of the origin, in which 0 is the only fixed point of the system. In this case, a Lyapunov function can be chosen as

$$V(\mathbf{x}) = \mathbf{f}^\top(\mathbf{x}) \, P \, \mathbf{f}(\mathbf{x}) \,.$$

Furthermore, if $\mathcal{D} = R^n$ and $V(\mathbf{x}) \to \infty$ as $\|\mathbf{x}\| \to \infty$, then this asymptotic stability is global.

Proof. First, since $\mathbf{x}^* = 0$ is a fixed point, $\mathbf{f}(0) = 0$, so

$$V(0) = \mathbf{f}^\top(0) \, P \, \mathbf{f}(0) = 0 \,.$$

Then, for any $\mathbf{x} \neq 0$ in the neighborhood $\mathcal{D}$, $\mathbf{f}(\mathbf{x}) \neq 0$, since 0 is the only fixed point of $\mathbf{f}$ in $\mathcal{D}$. Therefore, $V(\mathbf{x}) = \mathbf{f}^\top(\mathbf{x}) P \, \mathbf{f}(\mathbf{x}) > 0$ since P is positive definite. Also,

$$
\begin{aligned}
\dot{V}(\mathbf{x}) &= \left[\frac{\partial V}{\partial \mathbf{x}}\right] \dot{\mathbf{x}} \\
&= \left[\mathbf{f}^{\top}(\mathbf{x}) \left[\frac{\partial V}{\partial \mathbf{x}}\right]^{\top} P + \mathbf{f}^{\top}(\mathbf{x}) P \left[\frac{\partial V}{\partial \mathbf{x}}\right]\right] \mathbf{f}(\mathbf{x}) \\
&= \mathbf{f}^{\top}(\mathbf{x}) \left[J^{\top}(\mathbf{x}) P + P J(\mathbf{x})\right] \mathbf{f}(\mathbf{x}) \\
&= \mathbf{f}^{\top}(\mathbf{x}) \left[J^{\top}(\mathbf{x}) P + P J(\mathbf{x}) + Q\right] \mathbf{f}(\mathbf{x}) - \mathbf{f}^{\top}(\mathbf{x}) Q \, \mathbf{f}(\mathbf{x}) \\
&< 0 \qquad \text{for all } \mathbf{x} \neq 0.
\end{aligned}
$$

Finally, the global stability follows from Theorem 2.3. $\qquad \square$

Note that the above matrix Q is used to guarantee the matrix $[J^{\top}(\mathbf{x})P + PJ(\mathbf{x})]$ to always be negative definite. The reason is that, for some $\mathbf{x}$ in the neighborhood of the zero fixed point, the Jacobian $J(\mathbf{x})$ may be zero, so that $J^{\top}P + PJ = 0$ which, in turn, implies that $\dot{V}(\mathbf{x}) = 0$ (see the proof above). But, if $Q > 0$, then it is guaranteed that $\dot{V}(\mathbf{x}) < 0$ for all $\mathbf{x} \neq 0$. Usually, a convenient choice is $Q = I$, the identity matrix.

Similar stability criteria can be established for discrete-time systems. A discrete-time system is called *autonomous* if its system function $\mathbf{f}$ does not explicitly contain an independent discrete-time variable k; otherwise, it will be called *nonautonomous* and denoted as $\mathbf{f}_k$, $k = 1, 2, \dots$.

Two main results on discrete-time system stabilities are summarized as follows.

Theorem 2.5 (First Method of Lyapunov).
[For discrete-time autonomous systems]
Let $\mathbf{x}^* = 0$ be a fixed point of the discrete-time autonomous system

$$
\mathbf{x}_{k+1} = \mathbf{f}(\mathbf{x}_k), \tag{2.11}
$$

where $\mathbf{f} : \mathcal{D} \to R^n$ is continuously differentiable in a neighborhood of the origin, $\mathcal{D} \subseteq R^n$, and let $J = \left[\partial \mathbf{f}/\partial \mathbf{x}_k\right]_{\mathbf{x}_k = \mathbf{x}^* = 0}$ be its Jacobian evaluated at this fixed point. If all the eigenvalues of J are strictly less than one in absolute value, then the system is asymptotically stable about its zero fixed point.

Theorem 2.6 (Second Method of Lyapunov).
[For discrete-time nonautonomous systems]
Let $\mathbf{x}^* = 0$ be a fixed point of the nonautonomous system

$$
\mathbf{x}_{k+1} = \mathbf{f}_k(\mathbf{x}_k), \tag{2.12}
$$

where $\mathbf{f}_k : \mathcal{D} \to R^n$ *is continuously differentiable in a neighborhood of the origin,* $\mathcal{D} \subseteq R^n$. *Then the system* (2.12) *is globally (over the entire domain* $\mathcal{D}$*) and asymptotically stable about its zero fixed point, if there exists a scalar-valued function,* $V(\mathbf{x}_k, k)$, *defined on* $\mathcal{D}$ *and continuous in* $\mathbf{x}_k$, *such that*

(i) $V(0, k) = 0$ *for all* $k \geq k_0$;

(ii) $V(\mathbf{x}_k, k) > 0$ *for all* $\mathbf{x}_k \neq 0$ *in* $\mathcal{D}$ *and for all* $k \geq k_0$;

(iii) $\Delta V(\mathbf{x}_k, k) := V(\mathbf{x}_k, k) - V(\mathbf{x}_{k-1}, k-1) < 0$ *for all* $\mathbf{x}_k \neq 0$ *in* $\mathcal{D}$ *and for all* $k \geq k_0 + 1$;

(iv) $0 < W(||\mathbf{x}_k||) < V(\mathbf{x}_k, k)$ *for all* $k \geq k_0 + 1$, *where* $W(\cdot)$ *is a positive continuous function defined on* $\mathcal{D}$, *satisfying* $W(||0||) = 0$ *and* $\lim_{\tau \to \infty} W(\tau) = \infty$ *monotonically.*

As a special case, for discrete-time autonomous systems, Theorem 2.6 reduces to the following simple form.

Theorem 2.7 (Second Method of Lyapunov).

[For discrete-time autonomous systems]

Let $\mathbf{x}^* = 0$ *be a fixed point for the autonomous system* (2.11). *Then, the system is globally (over the entire domain* $\mathcal{D}$*) and asymptotically stable about this zero fixed point if there exists a scalar-valued function,* $V(\mathbf{x}_k)$, *defined on* $\mathcal{D}$ *and continuous in* $\mathbf{x}_k$, *such that*

(i) $V(0) = 0$;

(ii) $V(\mathbf{x}_k) > 0$ *for all* $\mathbf{x}_k \neq 0$ *in* $\mathcal{D}$;

(iii) $\Delta V(\mathbf{x}_k) := V(\mathbf{x}_k) - V(\mathbf{x}_{k-1}) < 0$ *for all* $\mathbf{x}_k \neq 0$ *in* $\mathcal{D}$;

(iv) $V(\mathbf{x}) \to \infty$ *as* $||\mathbf{x}|| \to \infty$.

To this end, it is important to emphasize that all the Lyapunov theorems stated above only offer *sufficient* conditions for asymptotic stability. On one hand, if no suitable Lyapunov function can be constructed to verify the conditions, one cannot say anything about the stability of the given system. On the other hand, usually more than one Lyapunov function may be constructed for the same system. Thus, for a given system, one choice of a Lyapunov function may yield a less conservative result (e.g. with a larger stability region) than other choices. Here, *stability region* refers to the set of system initial states starting from which the system trajectory will converge to the zero fixed point.

Nevertheless, there is a necessary condition in theory about the existence of a Lyapunov function.

Theorem 2.8 (Massera Inverse Theorem). *Suppose that the autonomous system* (2.9) *is asymptotically stable about its fixed point* $\mathbf{x}^*$ *and* $\mathbf{f}$ *is continuously differentiable with respect to* $\mathbf{x}$ *for all* $t \in [t_0, \infty)$. *Then, a Lyapunov function exists for this system.*

A combination of this theorem and Theorem 2.2 provides a necessary and sufficient condition for the asymptotic stability of a general autonomous system. For a general nonautonomous system, however, a necessary and sufficient condition of this type is possible but would be much more complicated.

Proof. See [Massera (1956)]. □

2.3 LaSalle Invariance Principle

First, the concepts of limit sets and invariant sets are introduced.

Definition 2.3. For a given dynamical system, a point $\mathbf{z}$ in the state space is said to be an ω-*limit point* of a trajectory $\mathbf{x}(t)$ of the system if for every open neighborhood $U_{\mathbf{z}}$ of $\mathbf{z}$, the trajectory of $\mathbf{x}(t)$ enters $U_{\mathbf{z}}$ at a finite time t. The set of all ω-limit points of $\mathbf{x}(t)$ is called the ω-*limit set* of $\mathbf{x}(t)$, which can be formulated as

$$\Omega^+(\mathbf{x}) = \Big\{ \mathbf{z} \in R^n \mid \text{ there exists } \{t_k\} \text{ with } t_k \to \infty$$
$$\text{such that } \mathbf{x}(t_k) \to \mathbf{z} \text{ as } k \to \infty \Big\}.$$

In the definition, ω is the last letter in the Greek alphabet, which is used to indicate the limiting process $t \to \infty$. Similarly, the α-*limit point* and the α-*limit set* are defined for the opposite limiting process as $t \to -\infty$, where α is the first letter of the Greek alphabet. Equivalently,

$$\Omega^-(\mathbf{x}) = \Big\{ \mathbf{z} \in R^n \mid \text{ there exists } \{t_k\} \text{ with } t_k \to -\infty$$
$$\text{such that } \mathbf{x}(t_k) \to \mathbf{z} \text{ as } k \to \infty \Big\}.$$

Simple examples of ω-limit points and ω-limit sets are fixed points and periodic trajectories (e.g. limit cycles), respectively, which however need not be stable.

Definition 2.4. An ω-limit set, $\Omega^+(\mathbf{x})$, is *attracting* if there exists an open neighborhood $U_{\Omega+}$ of $\Omega^+(\mathbf{x})$, such that when the trajectory of a system state enters $U_{\Omega+}$ at some instant $t_1 \geq t_0$, then this trajectory will approach

$\Omega^+(\mathbf{x})$ arbitrarily closely as $t \to \infty$. The *basin of attraction* of an attracting point is the union of all such open neighborhoods. An ω-limit set is *repelling* if the system trajectory always moves away from it.

Simple examples of attracting sets include stable fixed points and stable limit cycles, and repelling sets include those unstable ones.

Definition 2.5. A set $\mathcal{S} \subset R^n$ is said to be *forward (backward) invariant under function (or map)* $\mathbf{f}$ if, for all $\mathbf{x}(t) \in \mathcal{S}$, $\mathbf{f}(\mathbf{x}(t)) \subseteq \mathcal{S}$ for all $t \geq t_0$ ($t \leq t_0$).

As a note at this point, which will be further discussed in detail later, an *attractor* is defined to be the union of all those points in an attracting set that are invariant under the system function. In other words, an attractor is an ω-limit set, $\Omega^+(\mathbf{x})$, satisfying the property that all system trajectories near $\Omega^+(\mathbf{x})$ have $\Omega^+(\mathbf{x})$ as their ω-limit sets.

As another note, for a given map M in the discrete-time setting, an ω-limit set is similarly defined. Moreover, an ω-limit set is invariant under the map if $M(\Omega^+(\mathbf{x}_k)) = \Omega^+(\mathbf{x}_k)$ for all k; namely, starting from any point $\mathbf{x}_0$ in the set, the trajectory $\mathbf{x}_k = M^k(\mathbf{x}_0)$ will eventually return to the same set (but need not be the same point). Thus, an (invariant) ω-limit set embraces fixed points and periodic trajectories.

Theorem 2.9. *The ω-limit set of a finite state $\mathbf{x}$ of an autonomous system, $\dot{\mathbf{x}} = \mathbf{f}(\mathbf{x})$, is always invariant under $\mathbf{f}$.*

Proof. It amounts to showing that for any $\mathbf{z} \in \Omega^+(\mathbf{x})$, the system trajectory $\varphi_t(\mathbf{z}) \in \Omega^+(\mathbf{x})$.

By definition of $\Omega^+(\mathbf{x})$, there exists a sequence $\{t_k\}$ with $t_k \to \infty$ as $k \to \infty$, such that $\varphi_{t_k}(\mathbf{x}) \to \mathbf{z}$ as $k \to \infty$. Fix t and choose t_k sufficiently large so that $t + t_k > 0$. Then,

$$\varphi_{t+t_k}(\mathbf{x}) = \varphi_t(\varphi_{t_k}(\mathbf{x})) \to \varphi_t(\mathbf{z}) \qquad \text{as} \quad k \to \infty,$$

due to the time-invariant property of solutions of an autonomous system. Therefore, $\varphi_t(\mathbf{z}) \in \Omega^+(\mathbf{x})$, implying that $\Omega^+(\mathbf{x})$ is invariant under the map $\mathbf{f}$. $\square$

Now, consider once again the autonomous system (2.14) with a fixed point $\mathbf{x}^* = 0$.

Theorem 2.10 (LaSalle Invariance Principle). *Let $V(\mathbf{x})$ be a Lyapunov function of system (2.14), defined in a neighborhood $\mathcal{D}$ of the fixed* point $\mathbf{x}^* = 0$. Also, let $\varphi_t(\mathbf{x}_0)$ be a bounded solution trajectory of the system, with its initial state $\mathbf{x}_0$ and all its ω-limit points being confined within $\mathcal{D}$. Moreover, let

$$E = \left\{ \mathbf{x} \in \mathcal{D} \,\middle|\, \dot{V}(\mathbf{x}) = 0 \right\} \tag{2.13}$$

and $S \subset E$ be the largest invariant subset of E in the sense that if the initial state $\mathbf{x}_0 \in S$ then the entire trajectory $\varphi_t(\mathbf{x}_0) \subset S$ for all $t \geq t_0$.

Then, for any initial state $\mathbf{x}_0 \in \mathcal{D}$, the solution trajectory of the system satisfies

$$\varphi_t(\mathbf{x}_0) \to S \qquad \text{as} \quad t \to \infty.$$

Proof. Since $V(x)$ is a Lyapunov function, it is non-increasing in t and is bounded from below by zero. Hence,

$$V(\varphi_t(\mathbf{x}_0)) \to c \qquad (t \to \infty)$$

for a constant c depending only on $\mathbf{x}_0$.

Let $\Omega(\mathbf{x}_0)$ be the ω-limit set of $\mathbf{x}_0$. Then, $\Omega(\mathbf{x}_0)$ is an invariant set (Theorem 2.9) and belongs to $\mathcal{D}$ since the latter contains all limit points of $\varphi_t(\mathbf{x}_0)$.

Take a limit point of $\varphi_t(\mathbf{x}_0)$, $\mathbf{z} \in \Omega(\mathbf{x}_0)$. Then, there is an increasing subsequence $\{t_k\}$, with $t_k \to \infty$ as $k \to \infty$, such that $\varphi_{t_k}(\mathbf{x}_0) \to \mathbf{z}$ as $k \to \infty$. Therefore, by the continuity of V, $V(\mathbf{z}) = c$. Since $\mathbf{z}$ is arbitrarily chosen, this holds for all $\mathbf{z} \in \Omega(\mathbf{x}_0)$.

Note that, since $\Omega(\mathbf{x}_0)$ is invariant, if $\mathbf{z} \in \Omega(\mathbf{x}_0)$ then $\varphi_t(\mathbf{z}) \in \Omega(\mathbf{x}_0)$ for all $t \geq t_0$. Hence, $V(\mathbf{z}) = c$ for all $t \geq t_0$ and for all $\mathbf{z} \in \Omega(\mathbf{x}_0)$. This means

$$\dot{V}(\mathbf{z}) = 0 \qquad \text{for all } \mathbf{z} \in \Omega(\mathbf{x}_0).$$

Consequently, since S is the largest invariant subset of E, $\Omega(\mathbf{x}_0) \subseteq S \subseteq E$. But $\varphi_t(\mathbf{x}_0) \to \Omega(\mathbf{x}_0)$, so $\varphi_t(\mathbf{x}_0) \to S$, as $t \to \infty$. $\qquad\square$

This invariance principle is consistent with the Lyapunov theorems when they are applicable to a problem. Sometimes, when $\dot{V} = 0$ over a subset of the domain of V, a Lyapunov theorem is not easy to directly apply, but the LaSalle invariance principle may be convenient to use.

Example 2.11. Consider the system

$$\dot{x} = -x + \frac{1}{3}x^3 + y,$$
$$\dot{y} = -x.$$

The Lyapunov function $V = x^2 + y^2$ yields

$$\dot{V} = \frac{1}{2} x^2 \left(\frac{1}{3} x^2 - 1 \right) ,$$

which is negative for $x^2 < 3$ but is zero for $x = 0$ and $x^2 = 3$, regardless of the variable y. Thus, Lyapunov theorems do not seem to be applicable, at least not directly.

However, observe that the set E defined by (2.13) for this system has only three straight lines: $x = -\sqrt{3}$, $x = 0$, and $x = \sqrt{3}$. Note also that all trajectories intersecting the line $x = 0$ satisfy $\dot{y} = 0$ (the second equation) and so will not remain on the line $x = 0$ unless $y = 0$. This means that the largest invariant subset S containing the points with $x = 0$ is the only point $(0, 0)$.

Thus, it follows from Theorem 2.10 that, starting from any initial state located in a neighborhood of the point $(0, 0)$ bounded within the two stripes $x = \pm\sqrt{3}$, say located inside the disk

$$\mathcal{D} = \left\{ (x, y) \,|\, x^2 + y^2 < 3 \right\} ,$$

the solution trajectory will always be attracted to the point $(0, 0)$. This means that the system is (locally) asymptotically stable about its zero fixed point.

2.4   Some Instability Theorems

Once again, consider the general autonomous system (2.9), namely,

$$\dot{\mathbf{x}} = \mathbf{f}(\mathbf{x}), \qquad \mathbf{x}(t_0) = \mathbf{x}_0 \in R^n , \tag{2.14}$$

with a fixed point $\mathbf{x}^* = 0$. Sometimes, disproving the stability is easier than trying to find a Lyapunov function to prove it, since in this case a Lyapunov function can never be found. To disprove the stability, when the system indeed is unstable, the following instability theorems are useful.

Theorem 2.11 (Linear Instability Theorem).   *In system* (2.14), *let* $J = \left[\partial \mathbf{f} / \partial \mathbf{x} \right]_{\mathbf{x} = \mathbf{x}^* = 0}$ *be its Jacobian evaluated at* $\mathbf{x}^* = 0$. *If at least one of the eigenvalues of* J *has a positive real part, then the system is unstable about the fixed point* $\mathbf{x}^* = 0$.

For discrete-time systems, there is a similar result: *The discrete-time autonomous system*

$$\mathbf{x}_{k+1} = \mathbf{f}(\mathbf{x}_k), \qquad k = 0, 1, 2, \ldots ,$$

is unstable about its fixed point $\mathbf{x}^* = 0$ if at least one of the eigenvalues of the system Jacobian is larger than 1 in absolute value.

The following two negative theorems can be easily extended to nonautonomous systems in an obvious way.

Theorem 2.12 (General Instability Theorem). *For system* (2.14), *let $V(\mathbf{x})$ be a positive differential function defined on a neighborhood $\mathcal{D}$ of the origin, such that $V(0) = 0$ and in any arbitrarily small neighborhood of 0, there always exists an $\mathbf{x}_0$ which satisfies $V(\mathbf{x}_0) > 0$. If there is a closed subset $\Omega \subseteq \mathcal{D} \cap \mathcal{B}$ containing $\mathbf{x}_0$, where $\mathcal{B}$ is the unit ball of R^n, on which $V(\mathbf{x}) > 0$ and $\dot{V}(\mathbf{x}) > 0$, then the system is unstable about its zero fixed point.*

Proof. First, observe that since V is differentiable, it is continuous and so is bounded over any closed bounded set.

It can be shown that the trajectory $\mathbf{x}(t)$ of the system, starting from $\mathbf{x}(0) = \mathbf{x}_0$, must leave the subset Ω as time t evolves. To show this, note that as long as this trajectory $\mathbf{x}(t)$ is inside the subset Ω, one has $V(\mathbf{x}) > 0$. This is because $V(\mathbf{x})$ starts with the value $c := V(\mathbf{x}_0) > 0$, and $\dot{V}(\mathbf{x}) > 0$ by assumption. Let

$$\alpha = \inf \left\{ \dot{V}(\mathbf{x}) \,|\, \mathbf{x} \in \Omega \ \text{and} \ V(\mathbf{x}) \geq c > 0 \right\}.$$

Then, since Ω is a closed bounded set, $\alpha > 0$, so that

$$V(\mathbf{x}(t)) = V(\mathbf{x}_0) + \int_0^t \dot{V}(\mathbf{x}(\tau))\, d\tau \geq c + \int_0^t \alpha\, dt = c + \alpha\, t.$$

This inequality shows that $\mathbf{x}(t)$ cannot stay in Ω forever, because $V(\mathbf{x})$ is bounded on the closed subset Ω but now $V(\mathbf{x}) \geq c + \alpha t \to \infty$ as $t \to \infty$. Thus, $\mathbf{x}(t)$ must leave the set Ω.

Next, note that $\mathbf{x}(t)$ cannot leave Ω through the surface $V(\mathbf{x}) = 0$. This is because $V(\mathbf{x}(t)) \geq c > 0$, which implies that $\mathbf{x}(t)$ has to leave Ω through a path with the magnitude satisfying $\|\mathbf{x}\| \geq \|\mathbf{x}_0\|$ (since it starts with $\mathbf{x}(0) = \mathbf{x}_0$). According to the assumption, this $\mathbf{x}_0$ can be arbitrarily close to the origin, which implies that the origin is unstable: any trajectory starting from a point in an arbitrarily small neighborhood of the origin will eventually move away from it. $\square$

Example 2.12. Consider the system

$$\dot{x} = y + x\left(x^2 + y^4\right),$$
$$\dot{y} = -x + y\left(x^2 + y^4\right),$$

which has the fixed point $(0,0)$. The system Jacobian at the fixed point has a pair of imaginary eigenvalues, $\lambda_{1,2} = \pm j$, where $j = \sqrt{-1}$, so Theorem 2.1 is not applicable.

Note that the Lyapunov function $V = \frac{1}{2}\left(x^2 + y^2\right)$ leads to

$$\dot{V} = \left(x^2 + y^2\right)\left(x^2 + y^4\right) > 0, \qquad \text{for all} \quad (x,y) \neq (0,0).$$

Hence, this system is unstable about its zero fixed point by Theorem 2.12.

A variant of Theorem 2.12 is the following.

Theorem 2.13. *For system* (2.14), *let* $V(\mathbf{x})$ *be a positive differential function defined on a neighborhood* $\mathcal{D}$ *of the origin, such that* $V(0) = 0$ *and in any arbitrarily small neighborhood of 0, there always exists an* $\mathbf{x}_0$ *satisfying* $V(\mathbf{x}_0) > 0$. *If there is a closed subset* $\Omega \subseteq \mathcal{D} \cap \mathcal{B}$ *containing* $\mathbf{x}_0$, *where* $\mathcal{B}$ *is the unit ball of* R^n, *on which* $\dot{V}(\mathbf{x}) > aV(\mathbf{x})$ *for some constant* $a > 0$ *and for all* $\mathbf{x} \in \omega$, *then the system is unstable about its zero fixed point.*

Example 2.13. Consider the autonomous system

$$\dot{x} = x + 2y + xy^2,$$
$$\dot{y} = 2x + y - x^2 y.$$

Let $V(x,y) = x^2 - y^2$. Then, $V(x,y) > 0$ in the subset $\Omega = \left\{(x,y) \,\middle|\, |x| > |y|\right\}$, as shown in Fig. 2.8. Moreover,

$$\dot{V}(x,y) = 2x^2 - 2y^2 + 4x^2 y^2 = 2\,V(x,y) + 4x^2 y^2 \geq 2\,V(x,y),$$

for all $(x,y) \in \Omega$. Therefore, the system is unstable about its zero fixed point.

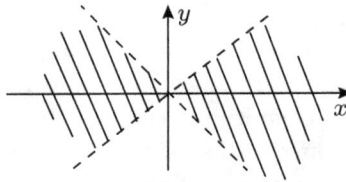

Fig. 2.8   The subset in which $|x| > |y|$.

Theorem 2.14 (Chetaev Instability Theorem).   *For system* (2.14), *let* $V(\mathbf{x})$ *be a positive and continuously differentiable function defined on* $\mathcal{D}$, *and let* Ω *be a subset of* $\mathcal{D}$, *containing the origin (i.e.* $0 \in \mathcal{D} \subseteq \Omega$). *If*

(i) $V(\mathbf{x}) > 0$ *and* $\dot{V}(\mathbf{x}) > 0$ *for all* $\mathbf{x} \neq 0$ *in* $\mathcal{D}$,

(ii) $V(\mathbf{x}) = 0$ *for all* $\mathbf{x}$ *on the boundary of* Ω,

then the system is unstable about its zero fixed point.

Proof. This instability theorem can be established by an argument similar to that given in the proof of Theorem 2.12. It is illustrated in Fig. 2.9, which graphically shows that if the theorem conditions are satisfied, then there is a gap within any neighborhood of the origin, so that it is possible for a system state trajectory to escape from the neighborhood of the origin along a path within this gap. □

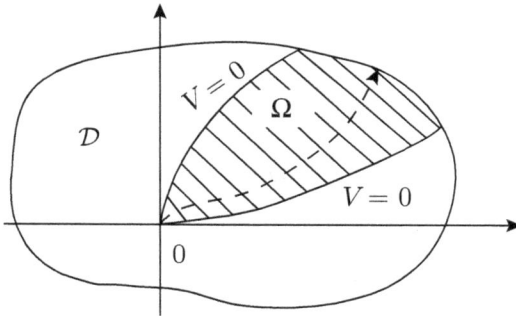

Fig. 2.9   Illustration of the Chetaev theorem.

Example 2.14. Consider the system

$$\dot{x} = x^2 + 2\,y^5\,,$$
$$\dot{y} = x\,y^2\,.$$

Choose the Lyapunov function

$$V = x^2 - y^4\,,$$

which is positive inside the region defined by

$$x = y^2 \qquad \text{and} \qquad x = -\,y^2\,.$$

Let $\mathcal{D}$ be the right-half plane and Ω be the shaded area $\mathcal{D}_0$ shown in Fig. 2.10. Clearly, $V = 0$ on the boundary of Ω, and $V > 0$ with $\dot{V} = 2x^3 > 0$ for all $(x, y) \in \mathcal{D}$. According to Theorem 2.14, this system is unstable about its zero fixed point.

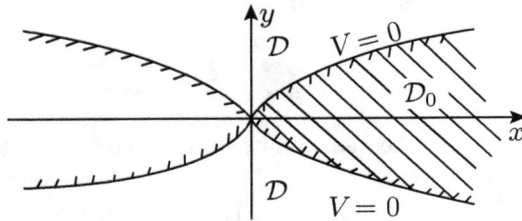

Fig. 2.10   The defining region of a Lyapunov function.

2.5   Construction of Lyapunov Functions

As seen from the previous subsections, Lyapunov functions play the central role in the Lyapunov stability theory, and how to find a simple working Lyapunov function is key to the determination of the stability or instability of a system with respect to its fixed point of interest.

There are many successful techniques for constructing a simple effective Lyapunov function. One natural choice is to use the total energy (kinetic energy and potential energy) of the system as the Lyapunov function, as illustrated by the following two examples.

Example 2.15. Consider the system

$$\dot{x} = y\,,$$
$$\dot{y} = -f(x)\,.$$

Let

$$V(x, y) = K + P = \frac{1}{2}y^2 + \int_0^x f(z)dz\,,$$

where K is the kinetic energy and P is the potential energy of the system. It can be easily verified that

$$\dot{V} = y\dot{y} + f(x)\dot{x} = -yf(x) + yf(x) = 0\,.$$

Hence, the system is stable in the sense of Lyapunov about its zero fixed point.

Example 2.16. Consider the dimensionless undamped pendulum (1.2) of unit mass, namely

$$\dot{x} = y\,,$$
$$\dot{y} = -\sin(x)\,,$$

where $x = \theta$ is the angular variable defined on $-\pi < x < \pi$.

It satisfies $\sin(0) = 0$ and $x\sin(x) > 0$ for $-\frac{\pi}{2} < x < \frac{\pi}{2}$. Let

$$V(x, y) = K + P = \frac{1}{2}y^2 + \int_0^x \sin(z)dz$$

$$= \frac{1}{2}y^2 + (1 - \cos(x)).$$

Since

$$\dot{V} = y\dot{y} + \sin(x)\dot{x} = -y\sin(x) + \sin(x)y = 0,$$

the pendulum is stable in the sense of Lyapunov about its zero fixed point for $-\frac{\pi}{2} < x = \theta < \frac{\pi}{2}$.

Another relatively simple and direct approach to constructing a Lyapunov function for an autonomous system is to apply Theorem 2.4, in which a Lyapunov function can be chosen as

$$V(\mathbf{x}) = \mathbf{f}^\top(\mathbf{x})P\mathbf{f}(\mathbf{x}),$$

where P and Q are two real positive definite constant matrices such that the matrix

$$J^\top(\mathbf{x})P + PJ(\mathbf{x}) + Q$$

is negative semi-definite for all $\mathbf{x} \neq 0$ in a neighborhood $\mathcal{D}$ of the zero fixed point. If these two matrices P and Q exist, then the given system is asymptotically stable about the zero fixed point in $\mathcal{D}$, and if furthermore $\mathcal{D} = R^n$ and $V(\mathbf{x}) \to \infty$ as $||\mathbf{x}|| \to \infty$, then this stability is global.

A limitation of this approach is that two constant matrices, P and Q, have to exist such that the matrix-valued function $J^\top(\mathbf{x})P + PJ(\mathbf{x}) + Q$ is negative semi-definite for *all* $\mathbf{x} \neq 0$ in $\mathcal{D}$. This existence is usually difficult or even impossible to prove. Therefore, more flexible methods are needed to construct a Lyapunov function in a general case.

To introduce one general method, let $V(\mathbf{x})$ be a Lyapunov function to be constructed, and let its gradient be defined by

$$\mathbf{g}^\top(\mathbf{x}) = \begin{bmatrix} g_1(\mathbf{x}) & \cdots & g_n(\mathbf{x}) \end{bmatrix} = \begin{bmatrix} \dfrac{\partial V(\mathbf{x})}{\partial x_1} & \cdots & \dfrac{\partial V(\mathbf{x})}{\partial x_n} \end{bmatrix}.$$

Then,

$$\dot{V}(\mathbf{x}) = \sum_{i=1}^n \frac{\partial V(\mathbf{x})}{\partial x_i}\dot{x}_i$$

$$= \begin{bmatrix} \dfrac{\partial V(\mathbf{x})}{\partial x_1} & \cdots & \dfrac{\partial V(\mathbf{x})}{\partial x_n} \end{bmatrix} \begin{bmatrix} f_1(\mathbf{x}) \\ \vdots \\ f_n(\mathbf{x}) \end{bmatrix}$$

$$= \mathbf{g}^\top(\mathbf{x})\mathbf{f}(\mathbf{x}).$$

Lemma 2.2. *The above* $\mathbf{g}(\mathbf{x})$ *is the gradient of a scalar-valued function if and only if*

$$\frac{\partial g_i}{\partial x_j} = \frac{\partial g_j}{\partial x_i}, \qquad \text{for all} \quad i, j = 1, \ldots, n.$$

Proof. If $\mathbf{g}(\mathbf{x})$ is the gradient of a scalar-valued function, $V(\mathbf{x})$, then

$$\mathbf{g}^{\top}(\mathbf{x}) = \begin{bmatrix} g_1(\mathbf{x}) & \cdots & g_n(\mathbf{x}) \end{bmatrix} = \begin{bmatrix} \dfrac{\partial V(\mathbf{x})}{\partial x_1} & \cdots & \dfrac{\partial V(\mathbf{x})}{\partial x_n} \end{bmatrix}.$$

It follows that

$$\frac{\partial g_i}{\partial x_j} = \frac{\partial}{\partial x_j} \frac{\partial V}{\partial x_i} = \frac{\partial^2 V}{\partial x_j \partial x_i}$$

$$= \frac{\partial^2 V}{\partial x_i \partial x_j} = \frac{\partial}{\partial x_i} \frac{\partial V}{\partial x_j} = \frac{\partial g_j}{\partial x_i},$$

for all $i, j = 1, \ldots, n$.

Conversely, if $\partial g_i / \partial x_j = \partial g_j / \partial x_i$ for all $i, j = 1, \ldots, n$, then a double integration gives

$$\int \int \frac{\partial g_i}{\partial x_j} \, dx_i dx_j = \int \int \frac{\partial g_j}{\partial x_i} \, dx_i dx_j,$$

or

$$\int g_i \, dx_i = \int g_j \, dx_j := V(\mathbf{x}),$$

which holds for all i and j, hence is independent of i and j, $i, j = 1, \ldots, n$. Thus,

$$\frac{\partial V}{\partial x_k} = \frac{\partial}{\partial x_k} \int g_k \, dx_k = g_k, \qquad k = 1, \ldots, n,$$

implying that $\mathbf{g}(\mathbf{x})$ is the gradient of the scalar-valued function $V(\mathbf{x})$. $\quad\square$

To construct the function $V(\mathbf{x})$, note that $V(\mathbf{x})$ can be calculated by the integration of its gradient:

$$V(\mathbf{x}) = \int_0^{\mathbf{x}} \sum_{i=1}^{n} \frac{\partial V(\mathbf{x})}{\partial x_i} \, dx_i = \int_0^{\mathbf{x}} \sum_{i=1}^{n} g_i(\mathbf{x}) \, dx_i.$$

Then, Lemma 2.2 shows that $\partial g_i / \partial x_j = \partial g_j / \partial x_i$, for all $i, j = 1, \ldots, n$. This means that the above integral is taken over a conservative field, so that the integral can be taken over any path joining 0 to the terminal state $\mathbf{x}$. In particular, this can be carried out along all the principal axes, namely,

$$V(\mathbf{x}) = \int_0^{x_1} g_1(\xi_1, 0, \ldots, 0) \, d\xi_1 + \int_0^{x_2} g_2(x_1, \xi_2, 0, \ldots, 0) \, d\xi_2$$

$$+ \cdots + \int_0^{x_n} g_n(x_1, \ldots, x_{n-1}, \xi_n) \, d\xi_n.$$

This formula can be used to construct a Lyapunov function, $V(\mathbf{x})$, for a given autonomous system, $\dot{\mathbf{x}} = \mathbf{f}(\mathbf{x})$, in a straightforward manner, as illustrated by the following example.

This approach to constructing Lyapunov functions is called the *variational gradient method*.

Example 2.17. Consider the damped pendulum system (1.1), namely,

$$\dot{x} = y$$
$$\dot{y} = -\frac{g}{\ell}\sin(x) - \frac{\kappa}{m}y\,,$$

where $x = \theta$ is the angular variable defined on $-\pi < x < \pi$.

One may start with a simple assumed form,

$$\mathbf{g}(x,y) = \begin{bmatrix} g_1(x,y) & g_2(x,y) \end{bmatrix}$$
$$= \begin{bmatrix} \alpha(x,y)\,x + \beta(x,y)\,y & \gamma(x,y)\,x + \delta(x,y)\,y \end{bmatrix},$$

where α, β, γ, δ are functions of x and y, and are to be determined so as to satisfy the condition $\partial g_1(x,y)/\partial y = \partial g_2(x,y)/\partial x$; that is,

$$\beta(x,y) + \frac{\partial\beta(x,y)}{\partial y}y + \frac{\partial\alpha(x,y)}{\partial y}x = \gamma(x,y) + \frac{\partial\gamma(x,y)}{\partial x}x + \frac{\partial\delta(x,y)}{\partial x}y\,. \quad \text{(a)}$$

On the other hand,

$$\dot{V}(x,y) = \frac{\partial V(x,y)}{\partial x}\dot{x} + \frac{\partial V(x,y)}{\partial y}\dot{y}$$

$$= \begin{bmatrix} g_1(x,y) & g_2(x,y) \end{bmatrix}\begin{bmatrix} f_1(x,y) \\ f_2(x,y) \end{bmatrix}$$

$$= \alpha(x,y)\,xy + \beta(x,y)\,y^2 - \frac{\kappa}{m}\gamma(x,y)\,xy$$

$$- \frac{g}{\ell}\gamma(x,y)\,x\sin(x) - \frac{\kappa}{m}\delta(x,y)\,y^2 - \frac{g}{\ell}\delta(x,y)\,y\sin(x)\,.$$

Then, some appropriate functions α, β, γ, δ should be chosen to simplify $\dot{V}(x,y)$, so that $V(x,y)$ can be easily found. For this purpose, one may proceed as follows:

(i) Cancel the crossed-product terms, by letting

$$\alpha(x,y)\,x - \frac{\kappa}{m}\gamma(x,y)\,x - \frac{g}{m}\delta(x,y)\,\sin(x) = 0\,, \quad \text{(b)}$$

so that

$$\dot{V}(x,y) = \begin{bmatrix} \beta(x,y) - \kappa\,m^{-1}\delta(x,y) \end{bmatrix}y^2 - g\,\ell^{-1}\gamma(x,y)\,x\sin(x)\,. \quad \text{(c)}$$

(ii) By inspection, use $\alpha = \alpha(x)$ and $\beta = \gamma = \delta = $ constant in equations (b) and (c), so that

$$g_1(x, y) = \alpha(x)\, x + \beta\, y, \quad g_2(x, y) = \gamma\, x + \delta\, y, \quad \text{and} \quad \beta = \gamma,$$

where the last equality follows from equation (a).

Thus, one has

$$
\begin{aligned}
V(x, y) &= \int_0^x g_1(\xi_1, 0)\, d\xi_1 + \int_0^y g_2(x, \xi_2)\, d\xi_2 \\
&= \int_0^x \alpha(\xi_1)\, \xi_1\, d\xi_1 + \int_0^y \left(\gamma\, x + \delta\, \xi_2 \right) d\xi_2 \\
&= \int_0^x \left(\frac{\kappa}{m} \xi_1 + \frac{g}{\ell}\, \delta\, \sin(\xi_1) \right) d\xi_1 + \int_0^y \left(\gamma\, x + \delta\, \xi_2 \right) d\xi_2 \\
&= \frac{\kappa}{2m}\, \gamma\, x^2 + \frac{g}{\ell}\, \delta - \frac{g}{\ell}\, \delta\, \cos(x) + \gamma\, xy + \frac{\delta}{2}\, y^2 \\
&= \frac{1}{2} \begin{bmatrix} x & y \end{bmatrix} \begin{bmatrix} (\kappa\gamma)/m & \gamma \\ \gamma & \delta \end{bmatrix} \begin{bmatrix} x \\ y \end{bmatrix} + \frac{g}{\ell}\, \delta\, [1 - \cos(x)].
\end{aligned}
$$

Now, to satisfy the requirements for a Lyapunov function, namely, $V(0, 0) = 0$ and $V(x, y) > 0$ with $\dot{V}(x, y) < 0$ for all $x \neq 0$ and $y \neq 0$, one may simply choose the constants γ and δ such that $0 < \gamma < \kappa\, m^{-1} \delta < \infty$. Then, it can be verified that this condition is sufficient to guarantee that $V(x, y)$ so constructed is a Lyapunov function. Consequently, the damped pendulum system is asymptotically stable about its zero fixed point in the domain of $-\pi < x = \theta < \pi$, consistent with the physics of the pendulum.

2.6   Stability Regions: Basins of Attraction

In the damped pendulum system (1.1), its fixed point $(\theta^*, \dot{\theta}^*) = (0, 0)$ is asymptotically stable, and any initial state satisfying the angular condition $-\pi < \theta_0 < \pi$ will eventually move to this fixed point at rest. In other words, the stability region, or basin of attraction, of the zero fixed point for this pendulum system is $(-\pi, \pi)$.

In general, for a nonautonomous system,

$$\dot{\mathbf{x}} = \mathbf{f}(\mathbf{x}, t), \qquad \mathbf{x}(t_0) = \mathbf{x}_0 \in R^n, \tag{2.15}$$

with a fixed point $\mathbf{x}^* = 0$, let $\varphi_t(\mathbf{x}_0)$ be its solution trajectory starting from $\mathbf{x}_0$. Suppose that this system is asymptotically stable about its zero fixed point. The interest here is to find out how large the region for the initial state $\mathbf{x}_0$ can be, within which one can guarantee that $\varphi_t(\mathbf{x}_0) \to 0$ as $t \to \infty$.

Definition 2.6. For system (2.15), the *region of stability*, or the *basin of attraction of its stable zero fixed point*, is the set

$$\mathcal{B} = \left\{ \mathbf{x}_0 \in R^n \mid \lim_{t \to \infty} \varphi_t(\mathbf{x}_0) = 0 \right\}.$$

For a stable linear system, $\dot{\mathbf{x}} = A\mathbf{x}$, it is clear that $\mathcal{B} = R^n$, the entire state space. For a nonlinear nonautonomous system, however, finding its exact basin of attraction is very difficult or even impossible in general. Therefore, one turns to find a (relatively conservative) estimate of the basin, to have a sense of "at least how large it would be". For this purpose, if one can find a Lyapunov function, $V(\mathbf{x})$, defined on a neighborhood of the zero fixed point, $\mathcal{D} \subseteq R^n$, and find a region

$$\Omega_c = \left\{ \mathbf{x} \in R^n \mid V(\mathbf{x}) \le c \right\}$$

where $c \ge 0$ is a constant, then all system trajectories starting from inside Ω_c will tend to zero as time evolves. In other words, the basin of attraction is at least as large as Ω_c.

Here, it is important to note that one may not simply use the defining domain $\mathcal{D}$ of the Lyapunov function as an estimate of the basin of attraction, as illustrated by the following example.

Example 2.18. Consider the autonomous system

$$\dot{x} = y,$$
$$\dot{y} = -x + \frac{1}{3}x^3 - y.$$

The Lyapunov function

$$V(x, y) = \frac{3}{4}x^2 - \frac{1}{12}x^4 + \frac{1}{2}xy + \frac{1}{2}y^2$$

satisfies

$$\dot{V}(x, y) = -\frac{1}{2}x^2\left(1 - \frac{1}{3}x^2\right) - \frac{1}{2}y^2,$$

so $V(x, y)$ is well defined on

$$\mathcal{D} = \left\{ (x, y) \in R^2 : \quad -\sqrt{3} < x < \sqrt{3} \right\}.$$

It is clear that $V(0,0) = 0$ and $V(x, y) > 0$ with $\dot{V}(x, y) < 0$ for all nonzero $(x, y) \in \mathcal{D}$. However, it can also be seen, from a computer plot shown in Fig. 2.11, that $\mathcal{D}$ is not a subset of $\mathcal{B}$, although they have a large overlapping region. In this example, $\mathcal{D}$ cannot be used as an estimate of $\mathcal{B}$.

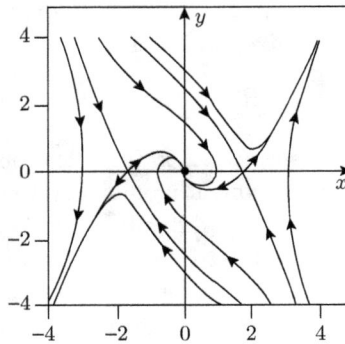

Fig. 2.11   Computer plot of the region of attraction.

Here, what is actually needed is to find a largest possible bounded subset, Ω_c, within the defining domain $\mathcal{D}$, for the given system. This may be
accomplished by following a general procedure, as illustrated by working
through this Example 2.18 as follows:

(1) Compute the system Jacobian evaluated at the zero fixed point:

$$J = \left[\frac{\partial \mathbf{f}(\mathbf{x})}{\partial \mathbf{x}}\right]_{\mathbf{x}=\mathbf{x}^*=0} = \begin{bmatrix} 0 & 1 \\ -1 & -1 \end{bmatrix}.$$

Make sure all its eigenvalues have negative real parts. Here, $\lambda_{1,2} = -0.5 \pm 1.5\,j$, where $j = \sqrt{-1}$.

(2) Choose a constant matrix $Q > 0$ (e.g. $Q = I$) and solve the Lyapunov
equation

$$J^\top P + PJ + Q = 0$$

for the positive definite matrix P. Then, find its eigenvalues:

$$P = \begin{bmatrix} 3/2 & 1/2 \\ 1/2 & 1 \end{bmatrix}, \qquad \lambda_{1,2} = \frac{5 \pm \sqrt{13}}{4}.$$

(3) Estimate a constant $c > 0$ using

$$c < \min_{||\mathbf{x}||=r} V(\mathbf{x}),$$

where r is a constant to be determined. In so doing, since $\mathbf{x}^\top P \mathbf{x}$ is a
Lyapunov function and

$$\mathbf{x}^\top P \mathbf{x} \geq \lambda_{\min}(P)\,||\mathbf{x}||^2,$$

one has the following estimate:

$$c < \lambda_{\min}(P)\,||\mathbf{x}||^2 = \frac{5 - \sqrt{13}}{4}\,r^2.$$

(4) Enlarge the estimate of the basin of attraction: Let $x = \rho \cos(\theta)$ and $y = \rho \sin(\theta)$, and compute

$$\dot{V}(\mathbf{x}) = \mathbf{x}^\top P \dot{\mathbf{x}} + \dot{\mathbf{x}}^\top P \mathbf{x}$$

$$= 2 \begin{bmatrix} x & y \end{bmatrix} \begin{bmatrix} 3/2 & 1/2 \\ 1/2 & 1 \end{bmatrix} \begin{bmatrix} y \\ -x + \frac{1}{3}x^3 - y \end{bmatrix}$$

$$= -x^2 + \frac{2}{3}x^3 y + \frac{1}{3}x^4 - y^2$$

$$= -\rho^2 + \frac{1}{3}\rho^4 \cos^3(\theta)\left(2\sin(\theta) + \cos(\theta)\right)$$

$$\leq -\rho^2 + \frac{1}{3}\rho^4 |2\sin(\theta) + \cos(\theta)|$$

$$\leq -\rho^2 + \frac{1}{3}\rho^4 \times 2.2361$$

$$< 0,$$

which gives $\rho^2 < 3/2.2361$, so that

$$c < \frac{5 - \sqrt{13}}{4} r^2 \leq \frac{5 - \sqrt{13}}{4} \rho^2 = \frac{5 - \sqrt{13}}{4} \cdot \frac{3}{2.2361} < 0.46771.$$

(5) An estimate of the basin of attraction for the system is finally obtained, as

$$\Omega_c = \left\{ \mathbf{x} \in R^2 \ : \ \mathbf{x}^\top \begin{bmatrix} 3/2 & 1/2 \\ 1/2 & 1 \end{bmatrix} \mathbf{x} \leq 0.46771 \right\}.$$

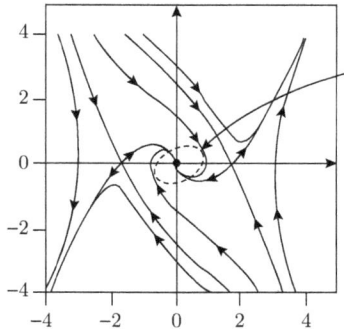

Fig. 2.12 Estimation of the region of attraction.

The resulting estimated basin of attraction, Ω_c, is an ellipse as shown in Fig. 2.12. Clearly, it is a rather conservative estimate as compared to the true basin given by the computer plot in Fig. 2.11 or Fig. 2.12.

Exercises

2.1 Verify which of the following systems is: (a) stable in the sense of Lyapunov only; (b) asymptotically but not exponentially stable; (c) exponentially stable:

$$\dot{x} = -\frac{x}{(1+t)^2}, \quad \dot{x} = -\frac{x}{1+t}, \quad \dot{x} = -\frac{x}{\sqrt{1+t}}, \quad \dot{x} = -(1+t)\,x\,.$$

[Hint: If necessary, these equations can be analytically solved by the standard method of separation of variables.]

2.2 Determine whether or not the following two systems have a stable fixed point. If so, explain whether the stability is (a) in the sense of Lyapunov only, (b) global, (c) asymptotic, (d) uniform, or (e) exponential.

$$\begin{bmatrix} \dot{x} \\ \dot{y} \end{bmatrix} = \begin{bmatrix} -1 & 2\sin(t) \\ 0 & -(t+1) \end{bmatrix} \begin{bmatrix} x \\ y \end{bmatrix} \quad \text{and} \quad \begin{bmatrix} \dot{x} \\ \dot{y} \end{bmatrix} = \begin{bmatrix} -1 & e^{2t} \\ 0 & -2 \end{bmatrix} \begin{bmatrix} \dot{x} \\ \dot{y} \end{bmatrix}.$$

2.3 * Prove that a linear time-varying system $\dot{x} = A(t)x$ is uniformly and asymptotically stable if and only if it is exponentially stable. Disprove this for nonlinear systems by the counterexample $\dot{x} = -x^3$ with solution $x(t) = x_0 / \left[1 + 2\,x_0^2(t - t_0)\right]^{1/2}$.

2.4 Consider the system

$$\dot{x} = 2x - 5y + x^2 - 4xy\,,$$
$$\dot{y} = 2x - 4y + 2x^2 - 3xy + 8y^2\,.$$

Use the Lyapunov first method to determine the stability of its zero fixed point. Try the Lyapunov function $V(\mathbf{x}) = 2x^2 - 6xy + 5y^2$, to see if it provides the same solution on the stability.

2.5 Consider the nonlinear system

$$\ddot{x} + 3\sqrt{1 + \dot{x}^2} + 2\,x = 0\,.$$

   (a) Find its fixed point on the x–$\dot{x}$ phase plane.
   (b) Determine the stability of the fixed point.
   (c) Determine the type (center, saddle, focus) of the fixed point.

2.6 Consider a time-varying damped pendulum described by

$$\dot{x} = y\,,$$
$$\dot{y} = -\frac{k(t)}{m}\,y - \frac{g}{\ell}\sin(x)\,,$$

where the time-varying dumping coefficient satisfies

$$a \le k(t) \le b \quad \text{and} \quad c \le \dot{k}(t) \le d$$

for some constants a, b, c, d. Under what conditions on a, b, c, d are its zero fixed point: (a) stable in the sense of Lyapunov, (b) asymptotically stable, (c) uniformly stable (if applicable), and (d) globally and asymptotically stable?

2.7 Analyze the stability of the following dynamical system:

$$\ddot{x} + 2a\,|x|\,\dot{x} + b\,x = c,$$

where $a > 0$, $b > 0$, and c are constants.

2.8 Consider the Volterra–Lotka system

$$\dot{x}(t) = a\,x(t) - b\,x(t)y(t),$$
$$\dot{y}(t) = b\,x(t)y(t) - c\,y(t),$$

where a, b, c are constants.

(a) Clearly, $(0, 0)$ is a fixed point of the system. Find the other one.
(b) Under what condition is $(0, 0)$ asymptotically stable?
(c) Linearize the system about $(0, 0)$, and then about another fixed point, respectively. Find the solutions of the two linearized systems.

2.9 Verify that the linear time-varying system

$$\dot{x} = \left[-1 - 9\cos^2(6t) + 12\sin(6t)\cos(6t)\right] x$$
$$\quad + \left[12\cos^2(6t) + 9\sin(6t)\cos(6t)\right] y,$$
$$\dot{y} = \left[-12\sin^2(6t) + 9\sin(6t)\cos(6t)\right] x$$
$$\quad - \left[1 + 9\sin^2(6t) + 12\sin(6t)\cos(6t)\right] y,$$

has eigenvalues $\lambda_1 = -1$ and $\lambda_2 = -10$, but is not even stable about its zero fixed point since its solution is (verify it):

$$x = c_1 e^{2t}(\cos(6t) + 2\sin(6t)) + c_2 e^{-13t}(\sin(6t) - 2\cos(6t)),$$
$$y = c_1 e^{2t}(2\cos(6t) - \sin(6t)) + c_2 e^{-13t}(2\sin(6t) + \cos(6t)).$$

2.10 Discuss the stability of the following system about its zero fixed point, by the Lyapunov first method and then by another method of your choice:

$$\begin{bmatrix} \dot{x} \\ \dot{y} \end{bmatrix} = \begin{bmatrix} \frac{17}{4} & -\frac{3}{4}e^t \\ \frac{199}{4}e^{-t} & -\frac{33}{4} \end{bmatrix} \begin{bmatrix} x \\ y \end{bmatrix}.$$

2.11 Given a linear system, $\dot{\mathbf{x}} = A\mathbf{x}$, with a constant matrix

$$A = \begin{bmatrix} 0 & 1 \\ a & b \end{bmatrix},$$

where $a < 0$ and $b < 0$ are constants. Prove its stability by constructing a Lyapunov function of the form $V(\mathbf{x}) = \mathbf{x}^\top P \mathbf{x}$ by using the positive definite matrix solution P of the Lyapunov equation

$$A^\top P + PA = -Q$$

in which $Q = \operatorname{diag}\{2, 2\}$.

2.12 Determine the stability of the fixed point of the following system:

$$\dot{x} = -2x + f(t)y,$$
$$\dot{y} = g(t)x - 2y,$$

where $|f(t)| \leq 1$ and $|g(t)| \leq 1$ for all $t \geq 0$.

2.13 Consider the nonhomogeneous system

$$\ddot{x}(t) + 2\dot{x}(t) = -f(x(t)),$$

where $f(\cdot)$ is a nonlinear differentiable function. Find the bounds a and b for the input, in the sense that $ax(t) \leq f(x(t)) \leq bx(t)$, such that within these bounds the system is asymptotically stable about its zero fixed point.

2.14 Use the Lyapunov function $V(\mathbf{x}) = x_1^2 + x_2^2$ to study the stability of the zero fixed point of the system

$$\dot{x} = x\left(\lambda^2 - x^2 - y^2\right) + y\left(x^2 + y^2 + \lambda^2\right),$$
$$\dot{y} = -x\left(\lambda^2 + x^2 + y^2\right) + y\left(\lambda^2 - x^2 - y^2\right),,$$

when (a) $\lambda = 0$ and (b) $\lambda \neq 0$.

2.15 Determine the stability of the zero fixed points of the following systems:

$$\dot{x} = -x - y,$$
$$\dot{y} = x - y^2,$$

and

$$\dot{x} = -x + y^2,$$
$$\dot{y} = -y.$$

2.16 Determine the stability of the zero fixed points of the following systems:

$$\dot{x} = x - y + xy,$$
$$\dot{y} = -y - y^2,$$

and

$$\dot{x} = x^2 + y^3,$$
$$\dot{y} = -y + x^3.$$

[Hint: Try $V(x, y) = (2x - y)^2 - y^2$ and $V(x, y) = x - y^2/2$, respectively.]

2.17 The following linear system is obviously stable in the sense of Lyapunov:

$$\dot{x} = y,$$
$$\dot{y} = -x,$$

since it is equivalent to $\ddot{x} = -x$, with solution $x(t) = c_1 \sin(t) + c_2 \cos(t)$. However, someone shows the following reasoning to argue that it is unstable even in the sense of Lyapunov. Explain why it is wrong:

Let $V(x,y) = (2x - y)^2 - y^2 = 4x^2 - 4xy$.

Then $V(x,y) > 0$ if $x > y$ (see Fig. 2.13 for a selected region Ω)

and $V(x,y) = 0$ on the boundaries $x = y$ and $x = 0$ of Ω

with $V(0,0) = 0$ and $(0,0) \in$ boundary of Ω.

However,

$$\dot{V}(x,y) = 8x\dot{x} - 4\dot{x}y - 4x\dot{y} = 8xy - 4y^2 - 4x(-x)$$
$$= 4(2x - y)y + 4x^2 > 0 \quad \text{inside } \Omega.$$

So, the equilibrium point (0,0) is unstable in the sense of Lyapunov.

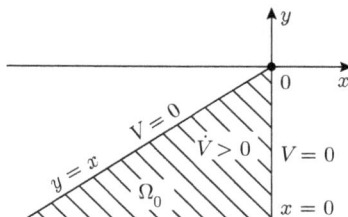

Fig. 2.13 Selected region of Ω.

2.18 Apply the LaSalle invariance principle to show that the Liénard equation

$$\ddot{x} + f(x)\,\dot{x} + g(x) = 0,$$

with $g(0) = 0$, is globally asymptotically stable about its zero fixed point, if the following conditions are satisfied:

$$f(x) > 0, \quad xg(x) > 0 \quad \text{for all } x \neq 0, \quad \text{and}$$

$$G(x) = \int_0^x g(z)\,dz \to \infty \quad (x \to \infty).$$

[Hint: Try $V(x,\dot{x}) = \frac{1}{2}y^2 + G(x)$, and note that the largest invariant set is $\mathcal{S} = \{(x,y)\colon f(x)y^2 = 0\}$, where $y = \dot{x}$.]

2.19 Apply the result of the previous exercise to the following systems:
$$\ddot{x} + x^2\dot{x} + x = 0$$
and to the van der Pol oscillator equation
$$\ddot{x} + \varepsilon(x^2 - 1)\dot{x} + x = 0 \qquad (\varepsilon < 0).$$

2.20 Consider the following autonomous system:
$$\dot{x} = y,$$
$$\dot{y} = -f(x),$$
satisfying $f(0) = 0$. Let the Lyapunov function be constructed in terms of the total system energy:
$$V(x, y) = \frac{1}{2}y^2 + \int_0^x f(z)dz.$$
Use this Lyapunov function to discuss the stability of the system with respect to the properties of f.

2.21 Use the variable gradient method to find a Lyapunov function for the system
$$\dot{x} = y,$$
$$\dot{y} = -4(x + y) - h(x + y),$$
where $h(\cdot)$ is a nonlinear function satisfying $h(0) = 0$ and $zh(z) \geq 0$ for all $|z| \leq 1$. [Hint: $V(\mathbf{x}) = 2x^2 + 2xy + y^2$ is one solution.]

2.22 Estimate the basin of attraction of the zero fixed point for the following system:
$$\dot{x} = -y,$$
$$\dot{y} = x + (x^2 - 1)y.$$
[Hint: Solve the Lyapunov equation to find the matrix $P > 0$ and then try the Lyapunov function $V(\mathbf{x}) = \mathbf{x}^\top P\mathbf{x}$. Note also that $|\cos^2\theta\sin\theta| \leq 0.3849$ and $|2\sin\theta - \cos\theta| \leq 2.2361$.]

2.23 For the following system:
$$\dot{x} = \left(\frac{1}{\sqrt{2}}z + \frac{\alpha}{2}\right)y - \beta x,$$
$$\dot{y} = \left(\frac{1}{\sqrt{2}}z - \frac{\alpha}{2}\right)x - \beta y,$$
$$\dot{z} = \sqrt{2}(1 - xy),$$
where $\alpha > 0$ and $\beta > 0$, show that in a *large* enough neighborhood of the origin there must be an attractor, and that this attractor is not the origin. [Hint: Construct a Lyapunov function to argue.]

2.24 Consider the following autonomous system:

$$\dot{x} = y \, ,$$
$$\dot{y} = -x + \left(1 - x^2 - y^2\right) y \, .$$

(a) Discuss the stability of the system about its zero fixed point.
(b) Find the limit cycle of the system.
(c) Use the Lyapunov method and the LaSalle principle to show that any system trajectory, if not starting from the origin, will converge to the limit cycle.

[Hint: Polar coordinates may be convenient to use.]

Chapter 3

Stabilities of Nonlinear Systems (II)

This chapter continues the study of nonlinear system stabilities. Specifically, the linear stability of nonlinear systems with periodic linearity will be discussed, followed by orbital stability, structural stability and total stability. Meanwhile, a useful toll of comparison principle will be introduced.

3.1 Linear Stability of Nonlinear Systems

The first method of Lyapunov provides a linear stability analysis for nonlinear autonomous systems. As can be seen from Theorems 2.1 and 2.11, the local stability and instability issues for autonomous systems are quite simple, which are similar to that of the familiar linear time-invariant systems. As has also been emphasized in the last chapter, the Lyapunov first method cannot be generally applied to nonautonomous systems. Thus, one may wonder to what extent the simple and efficient linear stability analysis methodology can be used for studying nonlinear nonautonomous systems.

In this section, the attention is first switched back onto the general nonautonomous system:

$$\dot{\mathbf{x}} = \mathbf{f}(\mathbf{x}, t), \qquad \mathbf{x}(t_0) = \mathbf{x}_0 \in R^n, \tag{3.1}$$

which is assumed to have a fixed point at $\mathbf{x}^* = 0$.

In system (3.1), a Taylor expansion of the function $\mathbf{f}$ about its zero fixed point gives

$$\dot{\mathbf{x}} = \mathbf{f}(\mathbf{x}, t) = J(t; t_0)\mathbf{x} + \mathbf{g}(\mathbf{x}, t), \tag{3.2}$$

where $J(t; t_0) = [\partial \mathbf{f}/\partial \mathbf{x}]_{\mathbf{x}=0}$ is the Jacobian and $\mathbf{g}(\mathbf{x}, t)$ is the residual of the expansion, which is assumed to satisfy

$$||\mathbf{g}(\mathbf{x}, t)|| \le a\, ||\mathbf{x}||^2 \qquad \text{for all} \quad t \in [t_0, \infty), \tag{3.3}$$

with constant $a > 0$. Recall from (2.5) that the solution of Eq. (3.2) is given by

$$\mathbf{x}(t) = \Phi(t, t_0)\,\mathbf{x}_0 + \int_{t_0}^{t} \Phi(t, \tau)\,\mathbf{g}(\mathbf{x}(\tau), \tau)\,d\tau\,, \tag{3.4}$$

where $\Phi(t, \tau)$ is the fundamental matrix of the system, defined in (2.4).

Theorem 3.1 (A General Linear Stability Theorem). *For the nonlinear nonautonomous system (3.2), if there are two positive constants, c and σ, such that*

$$\left\|\Phi(t, \tau)\right\| \le c\,e^{-\sigma(t-\tau)} \qquad \text{for all} \quad t_0 \le \tau \le t < \infty\,,$$

and if

$$\lim_{\|\mathbf{x}\| \to 0} \frac{\|\mathbf{g}(\mathbf{x}, t)\|}{\|\mathbf{x}\|} = 0$$

uniformly with respect to $t \in [t_0, \infty)$, then there are two positive constants, γ and δ, such that

$$\left\|\mathbf{x}(t)\right\| \le c\left\|\mathbf{x}_0\right\| e^{-\gamma(t-t_0)}$$

for all $\|\mathbf{x}_0\| \le \delta$ and all $t \in [t_0, \infty)$.

This result implies that under the theorem conditions, the system is locally, uniformly and exponentially stable about its zero fixed point. A proof of this theorem relies on the Gronwall Inequality, which is very useful in its own right (see, e.g. Example 3.3 below).

Lemma 3.1 (Gronwall Inequality). *Let $x(t)$, $y(t)$, and $z(t)$ be scalar-valued continuous functions defined on $[a, b]$, with $z(t) > 0$ and*

$$x(t) \le y(t) + \int_a^t z(\tau)x(\tau)\,d\tau\,, \qquad \text{for all} \quad t \in [a, b]\,.$$

Then, the following Gronwall Inequality:

$$x(t) \le y(t) + \int_a^t z(\tau)y(\tau)\,\exp\left\{\int_\tau^t z(s)ds\right\} d\tau\,, \tag{3.5}$$

holds for all $t \in [a, b]$.

Proof. Define

$$r(t) = \exp\left\{-\int_a^t z(\tau)d\tau\right\} \int_a^t z(\tau)x(\tau)d\tau\,, \quad t \in [a, b].$$

Then,

$$\dot{r}(t) = \left\{ x(t) - \int_a^t z(\tau) x(\tau) d\tau \right\} z(t) \exp\left\{ - \int_a^t z(\tau) d\tau \right\},$$

where, by the integral inequality in the assumed condition,

$$\left\{ x(t) - \int_a^t z(\tau) x(\tau) d\tau \right\} \le y(t).$$

Since $r(a) = 0$, integrating the above inequality from a to t gives

$$r(t) \le \int_a^t y(s) z(s) \exp\left\{ - \int_a^s z(\tau) d\tau \right\} ds.$$

Consequently,

$$\int_a^t z(\tau) x(\tau) d\tau = \exp\left\{ \int_a^t z(\tau) d\tau \right\} r(t)$$

$$\le \int_a^t y(\tau) z(\tau) \exp\left\{ \int_a^t z(s) ds - \int_a^\tau z(s) ds \right\} d\tau,$$

where $\int_a^t z(s) ds - \int_a^\tau z(s) ds = \int_\tau^t z(s) ds$.

Substituting this into the integral inequality in the assumed condition yields the Gronwall inequality. □

Now, a proof of Theorem 3.1 can be readily given.

Proof. Equation (3.4), or its slightly different form,

$$\mathbf{x}(t) = \Phi(t, t_0) \mathbf{x}_0 + \int_{t_0}^t \Phi(t, t_0) \Phi^{-1}(t_0, \tau) \mathbf{g}(\mathbf{x}(\tau), \tau) d\tau,$$

along with the bound on $\Phi(t, t_0)$ given in the theorem, yields

$$\|\mathbf{x}(t)\| \le c\,e^{-\sigma(t-t_0)} \|\mathbf{x}_0\| + \int_{t_0}^t c\,e^{-\sigma(t-\tau)} \|\mathbf{g}(\mathbf{x}(\tau), \tau)\| d\tau.$$

Due to the limiting condition given in the theorem, which means that $\mathbf{g}$ is a higher-order function of $\mathbf{x}$, one has $\|\mathbf{g}(\mathbf{x}, t)\| \le \beta \|\mathbf{x}\|$ for a constant $\beta > 0$, for (small enough) $\|\mathbf{x}\| < \beta/\alpha$ with $\alpha > 0$. It then follows from the Gronwall Inequality, with $x(t) = \|\mathbf{x}\| e^{\sigma t}$ therein, that

$$x(t) \le c\,e^{\sigma t_0} \|\mathbf{x}(t_0)\| + \beta c \int_{t_0}^t x(\tau) d\tau$$

which holds for small enough $\|\mathbf{x}\|$. Using again the Gronwall Inequality, but with constant functions $y = c\,e^{\sigma t_0} \|\mathbf{x}(t_0)\|$ and $z = \beta c$ at this time, yields

$$x(t) \le c\,e^{\sigma t_0} \|\mathbf{x}(t_0)\| \cdot e^{\beta c(t-t_0)}.$$

Therefore,

$$\|\mathbf{x}(t)\| \leq c\,\|\mathbf{x}(t_0)\| \cdot e^{-(\sigma - \beta c)\,(t - t_0)}\,.$$

Thus, by choosing $\beta < \sigma/c$ and letting $\gamma = \sigma - \beta c > 0$, and moreover restricting $\|\mathbf{x}(t_0)\|$ to be so small that $\|\mathbf{x}(t_0)\| < \delta := \beta/(\alpha c)$, the above inequality remains to be valid, with

$$\|\mathbf{x}(t)\| \leq c\,\|\mathbf{x}_0\|\,e^{-\gamma(t - t_0)}$$

which holds for all $\|\mathbf{x}_0\| \leq \delta$ and all $t \in [t_0, \infty)$. $\square$

Example 3.1. The simple nonlinear autonomous system

$$\dot{x} = -a\,x + x^3$$

is exponentially stable about its zero fixed point, if $a > 0$, in a small neighborhood of zero.

The stability of this system is obvious: the linear part of the system, $\dot{x} = -a\,x$ is exponentially stable, while the nonlinear part, x^3, is extremely small in $|x|$ near the origin. Therefore, the linear stability dominates the nonlinear instability in this system.

Note, however, that not all solutions of this system tend to the zero fixed point, which can be verified simply by sketching the phase portrait of its solution flow, or be understood from the fact that for large $|x|$, the term x^3 will dominate the linear part, $-a\,x$. Hence, starting from an initial point relatively far from the origin, the system trajectory will diverge.

Example 3.2. Consider the damped pendulum (1.1), now in the following form:

$$\begin{bmatrix} \dot{\theta} \\ \ddot{\theta} \end{bmatrix} = \begin{bmatrix} 0 & 1 \\ -\frac{g}{\ell} & -\frac{\kappa}{m} \end{bmatrix} \begin{bmatrix} \theta \\ \dot{\theta} \end{bmatrix} + \begin{bmatrix} 0 \\ \frac{g}{\ell}\,(\theta - \sin(\theta)) \end{bmatrix}.$$

It is easy to verify that the above constant system matrix is stable. Moreover,

$$\lim_{|\theta| \to 0} \frac{|\theta - \sin(\theta)|}{|\theta|} = 0\,.$$

Therefore, all the conditions of Theorem 3.1 are satisfied. The system is asymptotically stable about its zero fixed point.

If the following truncated expansion is used:

$$\left| \frac{g}{\ell}\,(\theta - \sin(\theta)) \right| \approx \frac{g}{\ell}\,\left| \frac{\theta^3}{3!} + \cdots \pm \frac{\theta^{2n+1}}{(2n+1)!} \right| \leq c\,\bigl(|\theta|^3 + \cdots + |\theta|^{2n+1} \bigr)$$

for some constant $c > 0$, then Theorem 3.1 is also applied to conclude that the zero fixed point of the damped pendulum is asymptotically stable.

Note that this example shows that the use of a more accurate approximate model may not make a radical difference in asymptotic stability analysis, as for the pendulum with a small enough angle here.

Example 3.3. Consider a perturbed linear time-invariant system,

$$\dot{\mathbf{x}}(t) = \left[A + \Delta_A(t) \right] \mathbf{x}(t), \qquad \mathbf{x}(t_0) = \mathbf{x}_0 \in R^n,$$

where the nominal matrix A is stable (i.e. all its eigenvalues have negative real parts). If the perturbation of the system matrix, $\Delta_A(t)$, is continuous and satisfies

$$||\Delta_A(t)|| < c \qquad \text{for all } t_0 \leq t < \infty$$

for a small enough constant $c > 0$, then the perturbed system will remain asymptotically stable about its zero fixed point.

To show this, however, the limiting condition stated in Theorem 3.1 cannot be verified. Thus, one has to resort to a direct application of the Gronwall Inequality (3.5).

The stability of matrix A implies that there are positive constants α and β such that $||\Phi(t, t_0)|| \leq \alpha\, e^{-(t-t_0)\beta}$ for all $t \geq t_0$, so that the solution

$$\mathbf{x}(t) = \Phi(t, t_0)\mathbf{x}_0 + \int_{t_0}^{t} \Phi(t, \tau)\Delta_A(\tau)\mathbf{x}(\tau)\, d\tau$$

satisfies

$$||\mathbf{x}(t)||\, e^{\beta(t-t_0)} \leq \alpha||\mathbf{x}_0|| + \int_{t_0}^{t} \alpha||\Delta_A(\tau)||\, ||\mathbf{x}(\tau)||\, e^{\beta(t-\tau)}\, d\tau$$

$$\leq \alpha||\mathbf{x}_0|| \exp\left\{ \alpha \int_{t_0}^{t} ||\Delta_A(\tau)||\, d\tau \right\}$$

$$\leq \alpha||\mathbf{x}_0||\, e^{\alpha c\,(t-t_0)},$$

where the Gronwall Inequality (3.5) has been used. It then follows that

$$||\mathbf{x}(t)|| \leq \alpha\,||\mathbf{x}_0||\, e^{(\alpha c - \beta)\,(t-t_0)}.$$

If c is small enough such that $\alpha c - \beta < 0$, then $||\mathbf{x}(t)|| \to 0$ as $t \to \infty$.

Note that the system $\dot{\mathbf{x}}(t) = \left[A + \Delta_A(t) \right] \mathbf{x}(t)$ in this example can be viewed as

$$\dot{\mathbf{x}}(t) = A\,\mathbf{x}(t) + \varepsilon\, \mathbf{g}(\mathbf{x}(t), t), \qquad \mathbf{x}(t_0) = \mathbf{x}_0 \in R^n,$$

where $\varepsilon > 0$ is a constant and $||\mathbf{g}|| \leq \delta$ uniformly for all $\mathbf{x} \in R^n$ for all $t \geq t_0$, with the constant $\delta > 0$ such that $\varepsilon\,\delta < c$ for a small enough $c > 0$. This means that a uniformly small (even nonlinear and time-varying) perturbation does not alter the stability of the linear part of the system.

Next, return to system (3.2). If the Jacobian $J(t;t_0) = J$ therein is a stable constant matrix, then the following simple criterion (as a special case of Example 2.6) is convenient to use.

Theorem 3.2. *Suppose that in system (3.2), the matrix $J(t;t_0) = J$ is a stable constant matrix and $\mathbf{g}(0,t) = 0$. Let P be a positive definite matrix solution of the Lyapunov equation*

$$PJ + J^\top P + Q = 0\,,$$

where Q is a positive definite constant matrix. If

$$||\mathbf{g}(\mathbf{x},t)|| \leq a\,||\mathbf{x}||$$

for a constant $a < \frac{1}{2}\,\lambda_{\max}(P)$ uniformly on $[t_0,\infty)$, then system (3.2) is globally, uniformly, and asymptotically stable about its zero fixed point.

A proof of this theorem was actually given in Example 2.6, and is a special case of the following slightly generalized result.

Theorem 3.3. *In system (3.2), suppose that both the matrix $J(t;t_0)$ and the function $\mathbf{g}(\mathbf{x},t)$ are continuous, and that there is a continuous non-negative function $\gamma(t)$ satisfying $\int_{t_0}^{\infty}\gamma(\tau)d\tau \leq a < \infty$ and $||\mathbf{g}(\mathbf{x},t)|| \leq \gamma(t)||\mathbf{x}(t)||$. Under these conditions, if $J(t;t_0)$ is uniformly stable (with all eigenvalues having a negative real part for all $t \geq t_0$), then the system (3.2) is locally, uniformly, and asymptotically stable about its zero fixed point.*

Proof. Note that the matrix $J(t;t_0)$ is uniformly stable. Thus, according to Exercise 3.6, there is a constant c such that $||\Phi(t,\tau)|| \leq c$ for all $t_0 \leq \tau \leq t < \infty$. For any $t^* \geq t_0$,

$$x(t) = \Phi(t,t^*)\mathbf{x}(t^*) + \int_{t^*}^{t} \Phi(t,\tau)\mathbf{g}(\mathbf{x}(\tau),\tau)\,d\tau\,,$$

where $t_0 \leq t^* \leq t < \infty$. It follows from the Gronwall inequality (3.5) that

$$||\mathbf{x}(t)|| \leq c\,||\mathbf{x}(t^*)|| + c\int_{t^*}^{t} \gamma(\tau)\,||\mathbf{x}(\tau)||\,d\tau$$

$$\leq c\,||\mathbf{x}(t^*)|| \exp\left\{ c\int_{t^*}^{t} \gamma(\tau)\,d\tau \right\}$$

$$\leq c\,||\mathbf{x}(t^*)|| \exp\left\{ c\int_{t^*}^{\infty} \gamma(\tau)\,d\tau \right\}$$

$$\leq c\,e^a\,||\mathbf{x}(t^*)||\,.$$

For any $\varepsilon > 0$, starting from an initial state satisfying $||\mathbf{x}(t^*)|| < \varepsilon e^{-a}/(2c)$, one has $||\mathbf{x}(t)|| < \varepsilon/2$. Then, an argument similar to that given in the proof of Theorem 3.1 shows that $||\mathbf{x}(t)|| < \varepsilon/2$ for all $t \geq t^*$. Therefore, $||\mathbf{x}(t)|| < \varepsilon$ for all $t_0 \leq t < \infty$. Hence, the system is stable in the sense of Laypunov about its zero fixed point.

Furthermore, since the matrix $J(t; t_0)$ is uniformly stable, $||\Phi(t, t_0)|| \to 0$ as $t \to \infty$. Thus, for any $\varepsilon > 0$ and any bounded $||\mathbf{x}_0||$, there is a $t^* > t_0$ such that $||\Phi(t, t_0)\mathbf{x}_0|| < \varepsilon$ for $t \geq t^*$, so that

$$||\mathbf{x}(t)|| \leq ||\Phi(t, t_0)\mathbf{x}_0|| + \int_{t_0}^{t} ||\Phi(t, \tau)|| \, ||\mathbf{g}(\mathbf{x}(\tau), \tau)|| \, d\tau$$

$$\leq \varepsilon + \int_{t_0}^{t} c\gamma(\tau) \, ||\mathbf{x}(\tau)|| \, d\tau$$

$$\leq \varepsilon \exp\left\{ c \int_{t_0}^{\infty} \gamma(\tau) \, d\tau \right\}$$

$$\leq \varepsilon \, e^{ac}, \quad t^* \leq t < \infty.$$

Since ε is arbitrary and e^{ac} does not depend on ε and t^*, it follows that $||\mathbf{x}(t)|| \to 0$ as $t \to \infty$ uniformly. $\qquad\square$

3.2 Linear Stability of Nonlinear Systems with Periodic Linearity

First, consider a linear time-varying system with a periodic coefficient matrix,

$$\dot{\mathbf{x}} = A(t)\,\mathbf{x}, \qquad \mathbf{x}(t_0) = \mathbf{x}_0 \in R^n, \qquad (3.6)$$

where $A(t + t_p) = A(t)$, with the fundamental period $t_p > 0$. It should be noted that a solution of this system may not be periodic, as can be seen from the following simple example.

Example 3.4. The system

$$\dot{x}(t) = [1 + \sin(t)]\, x(t)$$

has a 2π-periodic coefficient, but its nonzero solution is aperiodic:

$$x(t) = x_0 \, e^{t - \cos(t)}.$$

To further study system (3.6), let $\Phi(t, t_0)$ be the fundamental matrix associated with $A(t)$, which is defined by $\Phi(t, \tau) = \exp\left\{ \int_{\tau}^{t} A(\sigma)d\sigma \right\}$. Recall that all columns of $\Phi(t, t_0)$ are linearly independent solution vectors of (3.6), and the following is satisfied, $\dot{\Phi}(t, \tau) = A(t)\Phi(t, \tau)$ for $t_0 \leq \tau \leq t < \infty$.

For the t_p-periodic matrix $A(t + t_p) = A(t)$, it is clear that $\Phi(t + t_p) = \Phi(t)$, so that $\Phi(t + t_p, t_0)$ is another fundamental matrix associated with $A(t)$. Thus, the columns of $\Phi(t+t_p)$, being (linearly independent) solutions of (3.6), are linear combinations of those in $\Phi(t, t_0)$:

$$\phi_{ij}(t + t_p) = \sum_{\ell=1}^{n} c_{\ell j}\, \phi_{i\ell}\,,$$

where $\Phi = [\phi_{ij}]$ and $\{c_{ij}\}$ are constants, or

$$\Phi(t + t_p, t_0) = \Phi(t, t_0)\, C \tag{3.7}$$

with matrix $C = [c_{ij}]$. Since $\det[\Phi(t + t_p, t_0)] = \det[\Phi(t, t_0)] \cdot \det[C]$ and $\det[\Phi(t, t_0)] \neq 0$ for all $t \geq t_0$, one has $\det[C] \neq 0$.

It should be noted that all the eigenvalues of C are independent of the choice of the fundamental matrix $\Phi(t, t_0)$. To see this, let $\Phi_1(t, t_0)$ and $\Phi_2(t, t_0)$ be two fundamental matrices associated with $A(t)$. Then, since they consist of solutions of the same equation (3.6), there is a constant matrix, D, such that

$$\Phi_1(t, t_0) = \Phi_2(t, t_0)\, D\,,$$

and this D is nonsingular since both Φ_1 and Φ_2 are so. Thus, it follows that

$$\Phi_1(t + t_p; t_0) = \Phi_2(t + t_p; t_0)\, D = \Phi_2(t, t_0)\, CD$$
$$= \Phi_1(t, t_0) D^{-1} CD := \Phi_1(t, t_0)\, H\,.$$

This implies that H plays the role of C in Eq. (3.7). This also implies that H and C are similar nonsingular matrices, so they have the same eigenvalues. Therefore, a different choice of the fundamental matrix Φ in (3.7) only changes the form, but not the eigenvalues, of the constant matrix C therein.

Definition 3.1. The eigenvalues of the constant matrix C in Eq. (3.7) are called the *Floquet multipliers* of the system (3.6).

The importance of the Floquet multipliers can be appreciated from the observation that Eq. (3.7) implies

$$\Phi(t + nt_p, t_0) = \Phi(t, t_0)\, C^n\,, \qquad n = 1, 2, \ldots\,, \tag{3.8}$$

which, in turn, implies that the long-term dynamical behavior of a solution of system (3.6) is determined by the eigenvalues of C, namely, the Floquet multipliers.

Theorem 3.4 (Floquet Theorem).   *System* (3.6) *has at least one nonzero solution,* $\mathbf{x}(t)$, *and this solution satisfies*

$$\mathbf{x}(t + t_p) = \lambda\, \mathbf{x}(t)\,, \tag{3.9}$$

where λ *is a Floquet multiplier.*

Note that this solution, $\mathbf{x}(t)$, may not be periodic (unless $\lambda = 1$).

Proof. Let λ be an eigenvalue of C and let $\mathbf{v}$ be its associated eigenvector:

$$[C - \lambda I]\,\mathbf{v} = 0\,.$$

Let $\mathbf{x}(t) = \Phi(t, t_0)\mathbf{v}$. Then, it is straightforward to verify that $\mathbf{x}(t)$ is a nonzero solution of system (3.6). Moreover,

$$\mathbf{x}(t + t_p, t_0) = \Phi(t + t_p, t_0)\,\mathbf{v} = \Phi(t, t_0)\,C\,\mathbf{v} = \Phi(t, t_0)\,\lambda\,\mathbf{v} = \lambda\,\mathbf{x}(t)\,,$$

as claimed. □

The following example shows how to calculate the Floquet multipliers for a simple system.

Example 3.5. Consider the system

$$\begin{bmatrix} \dot{x} \\ \dot{y} \end{bmatrix} = \begin{bmatrix} 1 & 1 \\ 0 & \frac{\sin(t) + \cos(t)}{2 + \sin(t) - \cos(t)} \end{bmatrix} \begin{bmatrix} x \\ y \end{bmatrix},$$

with initial time $t_0 = 0$. This linear system can be easily solved, yielding

$$\begin{aligned} x(t) &= a\,e^t - b\left(2 + \sin(t)\right), \\ y(t) &= b\left(2 + \sin(t) - \cos(t)\right), \end{aligned} \tag{a}$$

with constants a and b determined by initial conditions. Hence, a corresponding fundamental matrix is

$$\Phi(t) = \begin{bmatrix} -2 - \sin(t) & e^t \\ 2 + \sin(t) - \cos(t) & 0 \end{bmatrix}.$$

Note that matrix $\Phi(t)$ is 2π-periodic up to a constant matrix multiplier, and the matrix C in (3.7) must satisfy $\Phi(t + 2\pi) = \Phi(t)C$ here, for all $t \geq 0$. Therefore, $\Phi(2\pi) = \Phi(0)C$, so that

$$C = \Phi^{-1}(0)\Phi(2\pi) = \begin{bmatrix} 1 & 0 \\ 0 & e^{2\pi} \end{bmatrix}.$$

Thus, the Floquet multipliers of the system are given by the eigenvalues of matrix C, which are $\lambda_1 = 1$ and $\lambda_2 = e^{2\pi}$.

In this example, $\lambda_1 = 1$ implies a 2π-periodic solution of the system (see (3.9)), which corresponds to $a = 0$ in the solution formula (a).

Now, in Eq. (3.7), let
$$R = \ln(C)/t_p\,,$$
where the principal value of the natural logarithm is taken, so $C = e^{t_p R}$. This constant matrix, R, is very useful. For example, it can be used to transform the periodic system (3.6) to be one having a constant coefficient matrix. Indeed, letting
$$\mathbf{x} = \left[\Phi(t, t_0)\, e^{-(t-t_0)R}\right] \mathbf{y}$$
changes the system (3.6) to
$$\dot{\mathbf{y}} = R\,\mathbf{y}\,.$$

Definition 3.2. In Definition 3.1, if for a Floquet multiplier λ, there is a real constant ρ such that
$$\lambda = e^{\rho\, t_p}\,,$$
then this ρ is called a *characteristic exponent* or *Floquet number* of the system (3.6).

In this definition, observe that
$$\rho = \ln(\lambda)/t_p\,,$$
where the principal value of the natural logarithm is taken. Since λ is an eigenvalue of the constant matrix C, ρ is an eigenvalue of the constant matrix R. Moreover, for the real part of ρ,
$$\mathrm{Re}\{\rho\} < 0 \qquad \Longleftrightarrow \qquad |\lambda| < 1\,,$$
which is useful for stability determination.

Now, consider a nonlinear nonautonomous system of the form
$$\dot{\mathbf{x}} = \mathbf{f}(\mathbf{x}, t) = J(t)\,\mathbf{x} + \mathbf{g}(\mathbf{x}, t)\,, \qquad \mathbf{x}(t_0) = \mathbf{x}_0 \in R^n\,, \tag{3.10}$$
where $\mathbf{g}(0, t) = 0$ and $J(t)$ is a p-periodic matrix, with $p > 0$:
$$J(t + p) = J(t) \qquad \text{for all } t \in [t_0, \infty).$$
Theorem 3.5 (Floquet Theorem). *In the system* (3.10), *assume that both* $\mathbf{g}(\mathbf{x}, t)$ *and* $\partial\mathbf{g}(\mathbf{x}, t)/\partial\mathbf{x}$ *are continuous in a bounded region* $\mathcal{D}$ *containing the origin. Assume also that*
$$\lim_{||\mathbf{x}||\to 0} \frac{||\mathbf{g}(\mathbf{x}, t)||}{||\mathbf{x}||} = 0$$
uniformly over $[t_0, \infty)$. *If the system Floquet multipliers satisfy*
$$|\lambda_i| < 1\,, \qquad i = 1, \ldots, n\,, \qquad \text{for all } t \in [t_0, \infty)\,, \tag{3.11}$$
then system (3.10) *is globally, uniformly, and asymptotically stable about its zero fixed point. In particular, this holds for the linear system* (3.10) *with* $\mathbf{g} = 0$ *therein.*

Proof. First, for the linear case with $\mathbf{g} = 0$, it follows from (3.9) and condition (3.11) that

$$\mathbf{x}(t_0 + nt_p) = \lambda_{\max}^n \mathbf{x}(t_0) \to 0 \qquad \text{as} \quad n \to \infty,$$

where $\lambda_{\max} = \max\{\lambda_i \colon i = 1, \ldots, n\}$. Then, for the general case, property (3.8) and condition (3.11) together imply that there exist constants $c > 0$ and $\sigma > 0$ such that

$$\|\Phi(t_0 + nt_p, t_0)\| \leq c e^{-\sigma n},$$

so that conditions similar to those in Theorem 3.1 are satisfied. Hence, the proof of Theorem 3.1 can be suitably modified for proving the present theorem. $\qquad \square$

One should compare this theorem with Theorem 3.1, which is for systems with an aperiodic coefficient matrix and so can achieve exponential stability. For systems with a periodic coefficient matrix, however, the best stability that one may expect is the asymptotic stability but not the exponential one. This is due to the periodicity of the system which, roughly, yields a stable solution converging to zero at the same rate as $c e^{-\sigma t} \sin(\omega t + \phi)$.

3.3 Comparison Principles

For large-scale and interconnected nonlinear systems, or systems described by differential inequalities rather than differential equations, the above stability criteria may not be directly applicable. In many such cases, the comparison principle is useful.

3.3.1 *Comparison Principle on the Plane*

3.3.1.1 *Comparison of zero points*

First, recall a concept from ordinary differential equations. A one-variable linear differential equation is said to be *homogeneous* if it satisfies

$$L(x) = 0,$$

where

$$L(\cdot) = \sum_{i=0}^{n} f_i(t) \frac{d^i}{dt^i}(\cdot),$$

in which f_i may be constant but not all zero, $i = 1, 2, \ldots, n$.

Example 3.6. The following second-order linear differential equations are homogeneous:

$$\sin(t)\ddot{x} + 2\dot{x} + x = 0\,, \qquad \ddot{x} + 2t\,\dot{x} + 3t^2 x = 0\,,$$

but the following are not:

$$\ddot{x} + 2\dot{x} + x = \sin(x)\,, \qquad \ddot{x} + 2t\,\dot{x} + 3t^2 x = 1\,.$$

Theorem 3.6 (Sturm Separation Theorem). *Consider a second-order homogeneous linear differential equation,*

$$\ddot{x} + c(t)x = 0\,,$$

defined on $[t_0, T]$, $T \leq \infty$, where $c(t)$ is a continuous function. Let $\phi_1(t)$ and $\phi_2(t)$ be its two fundamental solutions. Then, all zeros of the nontrivial solutions are isolated. Moreover, for any two successive zeros t_1 and t_2 of $\phi_1(t)$, solution $\phi_2(t)$ has exactly one zero located in (t_1, t_2).

Proof. Assume that the zeros of a nontrivial solution is not isolated. Then, since the zero set has a limit point, and the solution is a continuous function, the initial value problem has a zero solution, a contradiction.

Then, assume that the nontrivial solution $\phi_2(t)$ does not have zero in (t_1, t_2), then it does not have zero on $[t_1, t_2]$ since $\phi_1(t)$ and $\phi_2(t)$ are two fundamental solutions, hence are linearly independent. Thus, one can define a function

$$\psi(t) = \frac{\phi_1(t)}{\phi_2(t)}$$

on $[t_1, t_2]$. This function is continuous on $[t_1, t_2]$, where $\dot{\psi}(t)$ exists on (t_1, t_2), and $\psi(t_1) = \psi(t_2) = 0$. Consequently, by Rolle's Theorem in Calculus, there exists a point $p \in (t_1, t_2)$ such that $\dot{\psi}(p) = 0$. This implies that the Wronskian determinant is zero, but it is impossible because $\phi_1(t)$ and $\phi_2(t)$ are linearly independent.

Finally, assume that $\phi_2(t)$ has two zeros in (t_1, t_2), say $t_1 < t_3 < t_4 < t_2$. Then, similarly to the above argument, $\phi_1(t)$ will have a zero in (t_3, t_4), but this contradicts the fact that t_1 and t_2 are two successive zeros. Hence, $\phi_2(t)$ cannot have more than one zero in (t_1, t_2). $\qquad\square$

Theorem 3.7 (Sturm Comparison Theorem). *Consider two second-order homogeneous linear differential equations,*

$$\ddot{x} + c_1(t)x = 0\,,$$
$$\ddot{x} + c_2(t)x = 0\,,$$

defined on $[t_0, T]$, $T \leq \infty$, where the two coefficients are continuous functions satisfying $c_1(t) \leq c_2(t)$ on $[t_0, T]$. Let $\phi_1(t)$ and $\phi_2(t)$ be two nontrivial solutions of the two equations, respectively. Then, between any two successive zeros t_1 and t_2 of $\phi_1(t) = 0$, there exists at least one zero of $\phi_2(t)$ unless $c_1(t) \equiv c_2(t)$ on $[t_0, T]$.

Proof. Without loss of generality, assume that $\phi_1(t) > 0$ on (t_1, t_2); otherwise, consider $-\phi_1(t)$. Thus, one has $\dot{\phi}_1(t_1) > 0$ and $\dot{\phi}_1(t_2) < 0$.

Suppose, on the contrary, that $\phi_2(t)$ does not have zero on (t_1, t_2) and, likewise, without loss of generality assume that $\phi_2(t) > 0$ on (t_1, t_2).

Multiplying the first equation by $\phi_1(t)$ and the second by $\phi_2(t)$ and then subtracting the resultant equations yield

$$\ddot{\phi}_1\phi_2 - \phi_1\ddot{\phi}_2 + (c_1 - c_2)\phi_1\phi_2 = 0\,,$$

which gives

$$\frac{d}{dt}\left(\dot{\phi}_1\phi_2 - \phi_1\dot{\phi}_2\right) = -(c_1 - c_2)\phi_1\phi_2\,.$$

Integrating both sides from t_1 to t_2 leads to

$$\left.\left(\dot{\phi}_1\phi_2 - \phi_1\dot{\phi}_2\right)\right|_{t_1}^{t_2} = -\int_{t_1}^{t_2} [c_1(t) - c_2(t)]\phi_1(t)\phi_2(t)dt\,.$$

Note that the left-hand side is non-positive, while the right-hand side is strictly positive unless $c_1(t) \equiv c_2(t)$ on (t_1, t_2). Thus, for the case of $c_1(t) \not\equiv c_2(t)$, this is a contradiction. ∎

It is remarked that if $c_1(t) \equiv c_2(t)$ then the two equations become the same, so $\phi_1(t) \equiv \phi_2(t)$ on (t_1, t_2). In this case, the theorem does not exist.

Definition 3.3. The equation $\frac{d}{dt}\left(p(t)\dot{x}\right) + q(t)x = 0$ is said to be *oscillatory* on a time interval $[t_0, T]$, $T \leq \infty$, if there exists a nontrivial solution of the equation with infinitely many zeros on the time interval.

Example 3.7. The equation $\ddot{x} + x = 0$ is oscillatory on $[t_0, \infty)$. Indeed, it has a nontrivial solution $x(t) = \sin(t)$, which has infinitely many zeros on $[t_0, \infty)$.

In fact, this equation has a general solution $x(t) = c\sin(t)$, with a constant c, and every particular solution has infinitely many zeros on $[t_0, \infty)$.

It can be proved that the equation $\frac{d}{dt}(p(t)\dot{x}) + q(t)x = 0$ is oscillatory on a time interval $[t_0, T]$, $T \leq \infty$, if and only if all its solutions are so.

Thus, it follows that the oscillatory equation (system) $\frac{d}{dt}(p(t)\dot{x}) + q(t)x = 0$ is bounded, in the sense that all its solutions are bounded.

Note, however, that this system is not stable about its zero fixed point in the sense of Lyapunov, simply because no solution trajectory will stay nearby the zero fixed point forever.

3.3.1.2 *Comparison of functions*

For a more general setting than the above Sturm comparison theorem, there is an extension as follows.

Lemma 3.2 (Comparison Lemma). *Consider the 1-dimensional nonautonomous system*

$$\dot{x} = f(x, t), \qquad x(t_0) = x_0,$$

where $f(x, t)$ is locally Lipschitz in $x \in \mathcal{D} \subseteq R$ and continuous in $t \in [t_0, \infty)$. Let $y(t)$ be a continuously differentiable function on $[t_0, \infty)$, satisfying

$$\dot{y}(t) \leq f(y, t), \qquad y(t_0) \leq x(t_0),$$

for all $x \in \mathcal{D}$ and all $t \in [t_0, \infty)$. Then,

$$y(t) \leq x(t), \qquad \text{for all } t \in [t_0, \infty).$$

It can be easily seen that this lemma would be useful for determining the stability of one system from that of another.

3.3.2 *Comparison Principle in Higher-Dimensional Spaces*

Unlike the case of planar systems discussed above, comparing two entangled trajectories in a higher-dimensional space is very difficult and, in fact, only in some special situations such comparison is possible. Moreover, there are various forms of comparison principle in the higher-dimensional setting. Here, one particular form is presented to show some ideas beyond the comparison principle for higher-dimensional systems.

Consider the following two n-dimensional nonautonomous systems:

$$\dot{x}_i = f_i(x_1, x_2, \ldots, x_n, t), \qquad x_i(t_0) = x_i^0, \tag{3.12}$$

where f_i is continuous in $t \in [t_0, \infty)$, $i = 1, 2, \ldots, n$, and

$$\dot{y}_i = g_i(y_1, y_2, \ldots, y_n, t), \qquad y_i(t_0) = y_i^0, \tag{3.13}$$

where g_i is continuous in $t \in [t_0, \infty)$, $i = 1, 2, \ldots, n$, with $n \geq 2$.

Theorem 3.8 (Comparison Principle for Higher-Dimensional Systems). *Suppose that, for all* $i, j = 1, 2, \ldots, n$,

(i) $f_i(x_1, x_2, \ldots, x_n, t) \geq g_i(y_1, y_2, \ldots, y_n, t)$;
(ii) $x_i^0 \geq y_i^0$;
(iii) $\partial f_i / \partial x_j \geq 0$, $i \neq j$.

Then, $y_i \geq 0$ *for all* $t \geq t_0$ *and* $i = 1, 2, \ldots, n$, *implying that* $x_i \geq 0$, *for all* $t \geq t_0$ *and* $i = 1, 2, \ldots, n$.

Proof. See [Ważewski (1950)]. $\qquad\square$

3.4 Orbital Stability

The orbital stability differs from the Lyapunov stabilities in that it concerns with the stability of a system output (or state) trajectory under small external perturbations.

Let $\varphi_t(\mathbf{x}_0)$ be a p-periodic solution, $p > 0$, of the autonomous system

$$\dot{\mathbf{x}}(t) = \mathbf{f}(\mathbf{x}), \qquad \mathbf{x}(t_0) = \mathbf{x}_0 \in R^n, \tag{3.14}$$

and let Γ represent the closed orbit of $\varphi_t(\mathbf{x}_0)$ in the phase space, namely,

$$\Gamma = \left\{ \mathbf{y} \mid \mathbf{y} = \varphi_t(\mathbf{x}_0), \ \ 0 \leq t < p \right\}.$$

Definition 3.4. The p-periodic solution trajectory, $\varphi_t(\mathbf{x}_0)$, of the autonomous system (3.14) is said to be *orbitally stable* if, for any $\varepsilon > 0$, there exits a constant $\delta = \delta(\varepsilon) > 0$ such that for any $\mathbf{x}_0$ satisfying

$$d(\mathbf{x}_0, \Gamma) := \inf_{\mathbf{y} \in \Gamma} \|\mathbf{x}_0 - \mathbf{y}\| < \delta,$$

the solution of the system, $\varphi_t(\mathbf{x}_0)$, satisfies

$$d(\varphi_t(\mathbf{x}_0), \Gamma) < \varepsilon, \qquad \text{for all } t \geq t_0.$$

The orbital stability is visualized in Fig. 3.1.

Theorem 3.9 (Orbital Stability Theorem). *Let* $\mathbf{x}(t)$ *be a* p-*periodic solution of an autonomous system. Suppose that the system has Floquet multipliers* λ_i, *with* $\lambda_1 = 0$ *and* $|\lambda_i| < 1$, *for* $i = 2, \ldots, n$. *Then, this periodic solution* $\mathbf{x}(t)$ *is orbitally stable.*

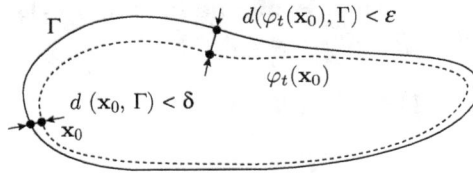

Fig. 3.1   Geometric meaning of the orbital stability.

Proof. See [Goddington and Levinson (1955)]: Chapter 13.2, Theorem 2.2.

$\square$

Example 3.8. For a simple example, a stable periodic solution, particularly a stable fixed point of a system, is orbitally stable. This is because all nearby trajectories approach it and, as such, it becomes a nearby orbit after a small perturbation and so will move back toward its original position (or stay nearby). On the contrary, unstable and saddle-node type of periodic orbits are orbitally unstable.

3.5   Structural Stability

Consider again the autonomous system (3.14).

To proceed, recall that a function or map is called a *diffeomorphism* if it is differentiable, invertible and its inverse is also differentiable.

Definition 3.5. Two system orbits are said to be *topologically equivalent* if there is a diffeomorphism that transfers one orbit to another. Two systems are said to be *topologically orbitally equivalent*, if their phase portraits are topologically equivalent.

If the dynamics of a system in the phase space changes radically, for instance by the appearance of a new fixed point or a new periodic orbit, due to small external perturbations, then the system is structurally unstable. To be more precise, consider the following set of functions:

$$\mathcal{S} = \left\{ \, \mathbf{g}(\mathbf{x}) \, \middle| \, ||\mathbf{g}(\mathbf{x})|| < \infty, \quad ||\partial \mathbf{g}(\mathbf{x})/\partial \mathbf{x}|| < \infty, \quad \forall \mathbf{x} \in R^n \, \right\}.$$

Definition 3.6. If, for any $\mathbf{g} \in \mathcal{S}$, there exists an $\varepsilon > 0$ such that the orbits of the two systems

$$\dot{\mathbf{x}} = \mathbf{f}(\mathbf{x}) \tag{3.15}$$

$$\dot{\mathbf{x}} = \mathbf{f}(\mathbf{x}) + \varepsilon \, \mathbf{g}(\mathbf{x}) \tag{3.16}$$

are topologically orbitally equivalent, then the autonomous system (3.15) is said to be *structurally stable*.

Example 3.9. System $\dot{x} = x^2$ is not structurally stable in any neighborhood of the origin. This is because, when the system is slightly perturbed, to become say $\dot{x} = x^2 + \varepsilon$, where $\varepsilon > 0$, then the resulting system has two equilibria, $x_1^* = \sqrt{\varepsilon}$ and $x_2^* = -\sqrt{\varepsilon}$, which has more numbers of fixed points than the original system that possesses only one, $x^* = 0$.

In contrast, $\dot{x} = x$ is structurally stable in any neighborhood of the origin, although the origin is an unstable fixed point of the system. To verify this, consider its perturbed system

$$\dot{x} = x + \varepsilon\, g(x),$$

where $g(x)$ is continuously differentiable and there is a constant $c > 0$ such that

$$|\,g(x)\,| < c \qquad \text{and} \qquad |\,g'(x)\,| < c, \qquad \text{for all } x \in R.$$

To find the fixed points of the perturbed system, let

$$f(x) := x + \varepsilon\, g(x) = 0$$

and choose ε such that $\varepsilon < 1/(2c)$. Then,

$$|\,\varepsilon g'(x)\,| < \varepsilon c < 1/2, \qquad \text{for all } x \in R,$$

so that

$$f'(x) = 1 + \varepsilon\, g'(x) > 1/2, \qquad \text{for all } x \in R.$$

This implies that the curve of the function $f(x)$ intersects the x-axis at exactly one single point. Hence, equation

$$\dot{x} = f(x) = x + \varepsilon\, g(x)$$

has exactly one solution, x^*. Since $g(x)$ is (uniformly) bounded, $|g(x)| < c$ for all $x \in R$. As $\varepsilon \to 0$, one has $\varepsilon g(x) \to 0$, so that $x^* \to 0$ as well. This means that the perturbed fixed point approaches the unperturbed one.

Furthermore, it must be shown that the perturbed fixed point is always unstable just like the unperturbed one. To do so, observe that the linearized equation of the perturbed system is

$$\begin{aligned}
\dot{x} = f(x) &\approx f(x_e) + f'(x)\big|_{x=x_e} (x - x_e) \\
&= 0 + \big[\,1 + \varepsilon\, g'(x_e)\,\big](x - x_e),
\end{aligned}$$

where x_e is the perturbed fixed point. Since

$$1 + \varepsilon\, g'(x_e) > 1 - \left| \varepsilon\, g'(x_e) \right| > 1 - \frac{1}{2} = \frac{1}{2} > 0\,,$$

the eigenvalue of the Jacobian $f'(x_e)$ is positive, implying that the perturbed system is unstable about its zero fixed point. Therefore, the unperturbed and perturbed systems are topologically orbitally equivalent; or, in other words, the unperturbed system is structurally stable about its zero fixed point.

Theorem 3.10 (Peixoto Structural Stability Theorem). *Consider a 2-dimensional autonomous system. Suppose that* **f** *is twice differentiable on a compact and connected subset* $\mathcal{D}$ *bounded by a simple closed curve,* Γ, *with an outward normal vector,* $\vec{n}$. *Assume that* $\mathbf{f} \cdot \vec{n} \neq 0$ *on* Γ. *Then, the system is structurally stable on* $\mathcal{D}$ *if and only if*

  (i) *all equilibria are hyperbolic;*
 (ii) *all periodic orbits are hyperbolic;*
(iii) *if* x *and* y *are hyperbolic saddles (probably,* $x = y$), *then* $W^s(x) \cap W^u(y)$ *is empty.*

Proof. See [Glendinning (1994)]: p. 92.   □

3.6 Total Stability: Stability under Persistent Disturbances

Consider a nonautonomous system and its perturbed version

$$\dot{\mathbf{x}} = \mathbf{f}(\mathbf{x}, t)\,, \qquad \mathbf{x}(t_0) = \mathbf{x}_0 \in R^n\,, \tag{3.17}$$

$$\dot{\mathbf{x}} = \mathbf{f}(\mathbf{x}, t) + \mathbf{h}(\mathbf{x}, t)\,, \tag{3.18}$$

where **f** is continuously differentiable, with $\mathbf{f}(0, t) = 0$, and **h** is a *persistent perturbation* in the sense defined below.

Definition 3.7. A function **h** is a *persistent perturbation* if, for any $\varepsilon > 0$, there are two positive constants, δ_1 and δ_2, such that

$$\|\mathbf{h}(\widetilde{\mathbf{x}}, t)\| < \delta_1\,, \qquad \forall t \in [t_0, \infty)\,,$$

and

$$\|\widetilde{\mathbf{x}}(t_0)\| < \delta_2$$

together implying $\|\widetilde{\mathbf{x}}(t)\| < \varepsilon$.

Definition 3.8. The zero fixed point of the unperturbed system (3.17) is said to be *totally stable*, if the persistently perturbed system (3.18) remains stable in the sense of Lyapunov.

As the next theorem states, all uniformly and asymptotically stable systems with persistent perturbations are totally stable, namely, a stable orbit starting from a neighborhood of another orbit will stay nearby.

Theorem 3.11 (Malkin Theorem). *If the unperturbed system (3.17) is uniformly and asymptotically stable about its zero fixed point, then it is totally stable, namely, the persistently perturbed system (3.18) remains stable in the sense of Lyapunov.*

Proof. See [Hoppensteadt (2000)]: p. 104.   □

Next, consider an autonomous system with persistent perturbations:

$$\dot{\mathbf{x}} = \mathbf{f}(\mathbf{x}) + \mathbf{h}(\mathbf{x}, t), \qquad \mathbf{x} \in R^n. \tag{3.19}$$

Theorem 3.12 (Perturbed Orbital Stability Theorem). *If $\varphi_t(\mathbf{x}_0)$ is an orbitally stable solution of the unperturbed autonomous system (3.19), with $\mathbf{h} = 0$ therein, then it is totally stable; that is, the perturbed system remains orbitally stable under persistent perturbations.*

Proof. See [Hoppensteadt (2000)]: p. 107.   □

Exercises

3.1 Compare and comment on the stabilities of the following two systems:
$$\dot{x} = -\,a\,\sin(x) \quad \text{and} \quad \dot{x} = -\,a\left(x - \frac{x^3}{3!} + \cdots \pm \frac{x^{2n+1}}{(2n+1)!}\right),$$
with constant $a > 0$.

3.2 Consider a linear time-varying system, $\dot{\mathbf{x}} = A(t)\,\mathbf{x}$, and let $\Phi(t,t_0)$ be its fundamental matrix. Show that this system is uniformly stable in the sense of Lyapunov about its zero fixed point if and only if $\|\Phi(t,\tau)\| \le c < \infty$ for a constant c and for all $t_0 \le \tau \le t < \infty$.

3.3 Consider a perturbed linear time-invariant system, $\dot{\mathbf{x}} = [A + \Delta_A(t)]\,\mathbf{x}$, with $\Delta_A(t)$ being continuous and satisfying $\int_{t_0}^{\infty} \|\Delta_A(\tau)\|\, d\tau \le c < \infty$. Show that this perturbed system is stable in the sense of Lyapunov about its zero fixed point.

3.4 * Consider the following nonlinear nonautonomous system:
$$\dot{\mathbf{x}} = \left[A + \Delta_A(t)\right]\mathbf{x} + \mathbf{g}(\mathbf{x},t),$$
which satisfies all conditions stated in Theorem 3.1. Assume, moreover, that $\Delta_A(t)$ is continuous with $\Delta_A(t) \to 0$ as $t \to \infty$. Show that this system is asymptotically stable about its zero fixed point. [Hint: Mimic the proof of Theorem 3.1 and Example 3.3.]

3.5 Determine the stability of the fixed point of the following system:
$$\dot{x} = -2\,x + f(t)\,y$$
$$\dot{y} = g(t)\,x - 2\,y,$$
where $|f(t)| \le 1/2$ and $|g(t)| \le 1/2$ for all $t \ge 0$.

3.6 For the following system:
$$\dot{x} = -\sin(2t)\,x + (\cos(2t) - 1)\,y$$
$$\dot{y} = (\cos(2t) + 1)\,x + \sin(2t)\,y,$$
find its Floquet multipliers and Floquet numbers.

3.7 Let $A(t)$ be a periodic matrix of period $\tau > 0$, and let $\{\lambda_i\}$ and $\{\rho_i\}$ be its Floquet multipliers and numbers, $i = 1, \ldots, n$, respectively. Verify that
$$\prod_{i=1}^{n} \lambda_i = \exp\left\{\int_{t}^{t+\tau} \text{trace}\,[A(s)]\, ds\right\}$$
and
$$\sum_{i=1}^{n} \rho_i = \frac{1}{\tau} \int_{t}^{t+\tau} \text{trace}\,[A(s)]\, ds.$$

3.8 Consider the following linear time-varying system with a periodic coefficient:

$$\dot{x} = (\sigma + \cos(t))\, x\,.$$

Verify that its Floquet multiplier is $\rho = \sigma$, and discuss the stability of the system about its zero fixed point with respect to the value of this multiplier.

3.9 Consider the predator-prey model

$$\dot{x} = -\alpha\, x + \beta\, x^2 - \gamma\, xy$$
$$\dot{y} = -\delta\, x + \varepsilon\, xy\,,$$

where all coefficients are positive constants. Discuss the stability of this system about its zero fixed point.

3.10 Consider the Mathieu equation

$$\ddot{x} + \big(a + b\,\cos(t)\big)x = 0\,.$$

Discuss the stability of its orbit about the zero fixed point in terms of the constants a and b.

3.11 A pendulum with mass m and weight-less string of length a hangs from a support that is constrained to move with vertical and horizontal displacements $\zeta(t)$ and $\eta(t)$, respectively. It can be verified that the motion equation of this pendulum is

$$a\,\ddot{\theta} + (g - \ddot{\zeta})\,\sin(\theta) + \ddot{\eta}\cos(\theta) = 0\,.$$

Assume that $\zeta(t) = \alpha\sin(\omega t)$ and $\eta(t) = \beta\sin(2\omega t)$, where $\omega = \sqrt{g/a}$. Show that the linearized equation, for small amplitudes, has a solution

$$\theta(t) = \frac{8\beta}{\alpha}\,\cos(\omega t)\,.$$

Discuss the stability of this solution.

3.12 Suppose that the Liénard equation

$$\ddot{x} + f(x)\,\dot{x} + g(x) = 0$$

has a t_p-periodic solution, $\varphi_t(x_0)$. Show that if

$$\int_0^{t_p} f(\varphi_\tau(x_0))\,d\tau > 0\,,$$

then this periodic solution is orbitally stable.

Chapter 4

Stabilities of Nonlinear Systems (III)

This chapter further studies the stabilities of nonlinear systems, introducing some analytic methods in the frequency domain, such as Lur'e systems formulation, harmonic balance approximation technique, bounded-input bounded-output stability concept and criteria, as well as the small gain theorem and the contraction mapping principle. All these are commonly used methodologies, very popular in engineering applications.

4.1   Lur'e Systems Formulated in the Frequency Domain

To motivate, consider a 1-dimensional nonlinear autonomous system,

$$\begin{cases} \dot{x} = f(x)\,, & -\infty < t < \infty\,, \\ x(-\infty) = 0\,, \end{cases}$$

where only causal signals are considered, so for simplicity the initial condition can be replaced by $x(0) = 0$, with $0 \leq t < \infty$. Using the unit-step function

$$g(t) = \begin{cases} 1 & t \geq 0\,, \\ 0 & t < 0\,, \end{cases}$$

the solution of this system can be written as

$$\begin{aligned} x(t) &= \int_{-\infty}^{t} f(x(\tau))\, d\tau \\ &= \int_{-\infty}^{\infty} g(t-\tau)\, f(x(\tau))\, d\tau \\ &= [g * f(x)]\,(t) \qquad \text{(convolution)}. \end{aligned}$$

The system can be implemented in the time domain as shown in Fig. 4.1.

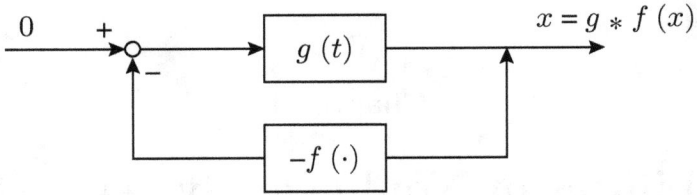

Fig. 4.1   Configuration of the 1-dimensional system.

In this configuration, if an "exponential signal" of the form e^{st} is input to the plant, represented by $g(\cdot)$, where s is a complex variable, then

$$g(t) * e^{st} = \int_{-\infty}^{\infty} g(\tau)\, e^{s(t-\tau)} d\tau$$

$$= \left[\int_{-\infty}^{\infty} g(\tau)\, e^{-s\tau} d\tau \right] e^{st}$$

$$:= G(s)\, e^{st} ,$$

in which $G(s)$ is actually the bilateral Laplace transform of $g(t)$ and serves as the *transfer function* of the linear plant represented by $g(\cdot)$. This is illustrated in Fig. 4.2.

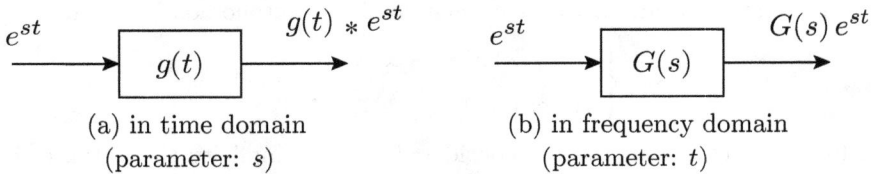

(a) in time domain   (b) in frequency domain
(parameter: s)   (parameter: t)

Fig. 4.2   Input–output relations of the plant $g(\cdot)$.

Now, to generalize, consider a higher-dimensional nonlinear autonomous system,

$$\begin{cases} \dot{\mathbf{x}} = \mathbf{f}(\mathbf{x}) , & -\infty < t < \infty , \\ \mathbf{x}(-\infty) = 0 , \end{cases}$$

where again only causal signals are considered, so the initial condition can be replaced by $\mathbf{x}(0) = 0$, with $0 \leq t < \infty$. Rewrite the system in the following form of *Lur'e system*:

$$\begin{cases} \dot{\mathbf{x}} = A\,\mathbf{x} + B\,\mathbf{h}(\mathbf{y}) \\ \mathbf{y} = C\,\mathbf{x} , \end{cases} \tag{4.1}$$

with initial conditions $\mathbf{x}(0) = 0$ or $\mathbf{y}(0) = 0$, where A, B, C are constant matrices, in which A is chosen by the user but B and C are determined from the given system (possibly, $B = C = I$), and $\mathbf{h}$ is a vector-valued nonlinear function generated through the reformulation. One specific and rather special example is

$$B = C = I, \qquad \mathbf{y} = \mathbf{x}, \qquad \mathbf{h} = \mathbf{f} - A,$$

with A chosen to possess a special property such as stability for convenience. Of course, there are other choices for the reformulation. The main purpose of this reformulation is to have a linear part, $A\mathbf{x}$, for the resulting system (4.1). In many applications, a given system may have already be given in the Lur'e form.

After the above reformulation, taking the Laplace transform $\mathcal{L}\{\cdot\}$ with zero initial conditions on the Lur'e system (4.1) yields

$$s\widehat{\mathbf{x}} = A\widehat{\mathbf{x}} + B\mathcal{L}\left\{\mathbf{h}(\mathbf{y})\right\},$$

or

$$\widehat{\mathbf{x}} = [sI - A]^{-1} B\mathcal{L}\left\{\mathbf{h}(\mathbf{y})\right\}.$$

where $\widehat{\mathbf{x}} = \mathcal{L}\{\mathbf{x}\}$. Consequently,

$$\widehat{\mathbf{y}} = C\widehat{\mathbf{x}} = C\left[sI - A\right]^{-1} B\mathcal{L}\left\{\mathbf{h}(\mathbf{y})\right\} := G(s)\,\mathcal{L}\{\mathbf{h}(\mathbf{y})\},$$

in which the system transfer matrix is

$$G(s) = C\left[sI - A\right]^{-1} B. \tag{4.2}$$

An implementation of the Lur'e system (4.1) is shown in Fig. 4.3, where both time-domain and frequency-domain notations are mixed for simplicity of expressions.

The Lur'e system shown in Fig. 4.3 is a closed-loop configuration, where the feedback loop is usually considered as a "controller." Thus, this system is sometimes written in the following equivalent control form:

$$\begin{cases} \dot{\mathbf{x}} = A\mathbf{x} + B\mathbf{u} \\ \mathbf{y} = C\mathbf{x} \\ \mathbf{u} = \mathbf{h}(\mathbf{y}). \end{cases} \tag{4.3}$$

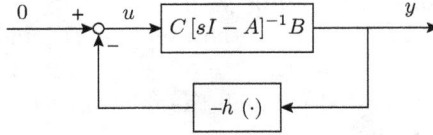

Fig. 4.3 Configuration of the Lur'e system.

4.2 Absolute Stability and Frequency-Domain Criteria

This section introduces the concept of absolute stability about the Lur'e system and derives some frequency-domain criteria for this type of stability.

4.2.1 *Background and Motivation*

To provide some background information and motivation, consider a single-input single-output (SISO) Lur'e system,

$$\begin{cases} \dot{\mathbf{x}} = A\mathbf{x} + \mathbf{b}\,u \\ y = \mathbf{c}^\top \mathbf{x} \\ u = h(y)\,. \end{cases} \tag{4.4}$$

Assume that $h(0) = 0$. Then, $\mathbf{x}^* = 0$ is a fixed point of the system. The transfer function of its linear part is given by $G(s) = \mathbf{c}^\top [sI - A]^{-1}\mathbf{b}$. Let $G_r(j\,\omega) = G(j\,\omega) + G(-j\,\omega)$ be the real part of $G(j\,\omega)$, and assume that

(i) $G(s)$ is stable and satisfies $G_r(j\,\omega) \geq 0$;
(ii) $u(t) = -g(\mathbf{x})\,y(t)$ with $g(\mathbf{x}) \geq 0$.

Based on condition (i), one can construct a rational function (as another transfer function), $R(s)$, and find a constant vector, $\mathbf{r}$, such that

$$R(s)R(-s) = G(s) + G(-s) \tag{4.5}$$

with

$$R(s) = \mathbf{r}^\top [sI - A]^{-1}\mathbf{b}\,. \tag{4.6}$$

This is known as *spectral factorization* in matrix theory.

Then, consider the Lyapunov equation

$$PA + A^\top P = -\mathbf{r}\,\mathbf{r}^\top$$

with a positive definite constant matrix, P. For

$$P\,[sI - A] + [-sI - A^\top]\,P = \mathbf{r}\,\mathbf{r}^\top,$$

multiplying $[sI - A]^{-1}\mathbf{b}$ to its right and $\mathbf{b}^\top[-sI - A^\top]^{-1}$ to its left yields

$$\mathbf{b}^\top P[-sI - A]^{-1}\mathbf{b} + \mathbf{b}^\top[-sI - A^\top]^{-1}P\mathbf{b} = R(-s)R(s). \qquad (4.7)$$

Thus, one can identify $\mathbf{c} = P\mathbf{b}$ by comparing (4.5) and (4.7).

To this end, the Lyapunov function $V(\mathbf{x}) = \mathbf{x}^\top P\mathbf{x}$ yields

$$\begin{aligned}\dot{V}(\mathbf{x}) &= -\mathbf{x}^\top \mathbf{r}\,\mathbf{r}^\top \mathbf{x} - 2\,\mathbf{x}^\top \mathbf{c}\,g(\mathbf{x})\,\mathbf{c}^\top \mathbf{x} \\ &= -\left(\mathbf{r}^\top \mathbf{x}\right)^2 - 2\left(\mathbf{c}^\top \mathbf{x}\right)^2 g(\mathbf{x}) \le 0.\end{aligned}$$

This implies that system (4.4) is stable in the sense of Lyapunov about its fixed point $\mathbf{x} = 0$.

Note that this stability can be strengthened to be asymptotic if conditions (i) and (ii) above are modified to be as follows:

(i)' $G(s)$ is stable and satisfies $G_r(j\,\omega) > 0$;

(ii)'' $u(t) = -g(\mathbf{x})\,y(t)$ with $g(\mathbf{x}) > 0$.

This is left to be further discussed below, where some even more convenient stability criteria will be derived.

Now, observe that, in many cases, either condition (i) or (ii) may not be satisfied by the given system (4.4). Nevertheless, if the function $g(\mathbf{x})$ satisfies the following weaker condition:

$$\alpha \le g(\mathbf{x}) < \beta \qquad \text{for all } \mathbf{x} \in R^n, \qquad (4.8)$$

for some constants α and β, then one may try to change the variables as follows:

$$u = \beta v - \alpha z \qquad \text{and} \qquad y = z - v,$$

so that

$$v(t) = -\tilde{g}(\mathbf{x})\,z(t)$$

where

$$\tilde{g}(\mathbf{x}) = \frac{g(\mathbf{x}) - \alpha}{\beta - g(\mathbf{x})} \ge 0,$$

which satisfies condition (ii). In this case, the new transfer function becomes

$$\widetilde{G}(s) = \frac{1 + \beta\,G(s)}{1 + \alpha\,G(s)}.$$

If this new transfer function satisfies condition (i), then all of the above analysis and results remain valid. This motivates the study of the stability problem under condition (4.8) below in the rest of the section.

4.2.2 SISO Lur'e Systems

First, single-input single-output (SISO) Lur'e systems of the form (4.4) are discussed, namely,

$$\begin{cases} \dot{\mathbf{x}} = A\mathbf{x} + \mathbf{b}\,u \\ y = \mathbf{c}^\top \mathbf{x} \\ u = h(y)\,. \end{cases} \tag{4.9}$$

Assume that $h(0) = 0$, so $\mathbf{x}^* = 0$ is a fixed point of the system.

In light of condition (4.8), the following *sector condition* is imposed.

Sector Condition: The Lur'e system (4.9) is said to satisfy the *local (global) sector condition* on the nonlinear function $h(\cdot)$, if there exist two constants, $\alpha \le \beta$, such that

(i) local sector condition:

$$\alpha\,y^2(t) \le y(t)\,h(y(t)) \le \beta\,y^2(t)\,, \tag{4.10}$$

for all $-\infty < y_m \le y(t) \le y_M < \infty$ and $t \in [t_0, \infty)$;

(ii) global sector condition:

$$\alpha\,y^2(t) \le y(t)\,h(y(t)) \le \beta\,y^2(t)\,, \tag{4.11}$$

for all $-\infty < y(t) < \infty$ and $t \in [t_0, \infty)$.

Here, $[\alpha, \beta]$ is called a *sector* for the nonlinear function $h(\cdot)$. Moreover, system (4.9) is said to be *absolutely stable within the sector* $[\alpha, \beta]$ if the system is globally asymptotically stable about its fixed point $\mathbf{x}^* = 0$ for any nonlinear function $h(\cdot)$ satisfying the global sector condition (ii).

The above local and global sector conditions are visualized in Figs. 4.4(a) and 4.4(b), respectively. It is easy to see that if a system satisfies the global sector condition then it also satisfies the local sector condition, but the converse may not be true.

As a historical remark, Aizerman made a conjecture in 1940s that if the Lur'e system is stable about its zero fixed point for all linear system approximations with corresponding constant slope κ satisfying $\alpha \le \kappa \le \beta$ as shown in Fig. 4.4, then the original nonlinear Lur'e system would be stable. In 1957, Kalman modified this to be an even stronger conjecture that if the system is stable for all linear system approximations with $\alpha_0 \le dh(y)/dy \le \beta_0$ for some $\alpha_0 \le \alpha$ and $\beta_0 \le \beta$, then the original nonlinear system would be stable. It is now known that they both are false because

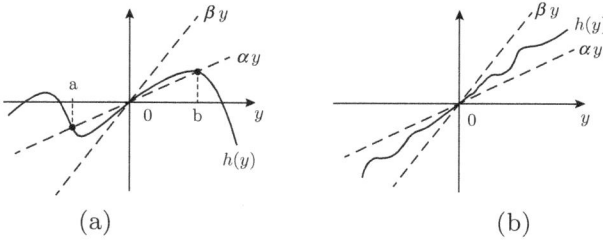

Fig. 4.4   Local and global sector conditions.

some counterexamples were found lately. One simple counterexample is given in Exercise 4.4 (see [Hahn (1967)] for another counterexample).

The following simple example shows, in some specific cases, how to determine a sector for a given system.

Example 4.1. The linear system

$$\begin{cases} \dot{x} = a\,x + b\,u \\ y = c\,x \\ u = h(y) = y \end{cases}$$

satisfies the global sector condition with $\alpha = \beta = 1$.

In this system, if the controller is changed to be

$$u = h(y) = 2\,y^2\,,$$

then the system becomes nonlinear but satisfies

$$-2\,y^2(t) \le y(t) \cdot 2\,y^2(t) \le 2\,y^2(t)$$

for all $-1 \le y(t) \le 1$ and $t \in [t_0, \infty)$. In this case, the system satisfies the local sector condition with $a = -1$ and $b = 1$, over the sector $[\alpha, \beta] = [-2, 2]$.

Theorem 4.1 (Popov Criterion). *Suppose that the SISO Lur'e system (4.9) satisfies the following conditions:*

(i) *A is stable and $\{A, \mathbf{b}\}$ is controllable;*
(ii) *the system satisfies the global sector condition with $\alpha = 0$ therein;*
(iii) *for any $\varepsilon > 0$, there is a constant $\gamma > 0$ such that*

$$\mathrm{Re}\left\{ (1 + j\,\gamma\,\omega)\, G(j\,\omega) \right\} + \frac{1}{\beta} \ge \varepsilon \qquad \text{for all} \quad \omega \ge 0, \qquad (4.12)$$

where $G(s)$ is the transfer function defined by (4.2), and $\mathrm{Re}\{\cdot\}$ denotes the real part of a complex number (or function). Then, the system is globally asymptotically stable about its fixed point $\mathbf{x}^* = 0$ within the sector.

Proof. See [Khalil (1996)]: pp. 419–421.   □

The Popov criterion has the following geometric meaning. Separate the complex function $G(s)$ into its real and imaginary parts:

$$G(j\,\omega) = G_r(\omega) + j\,G_i(\omega),$$

where

$$G_r(\omega) = \tfrac{1}{2}\left[G(j\,\omega) + G(-j\,\omega)\right],$$
$$G_i(\omega) = \tfrac{1}{2j}\left[G(j\,\omega) - G(-j\,\omega)\right].$$

Then, rewrite condition (iii) as

$$\frac{1}{\beta} > -G_r(\omega) + \gamma\,\omega\,G_i(\omega) \qquad \text{for all } \omega \geq 0.$$

Thus, the graphical situation of the Popov criterion as shown in Fig. 4.5 implies the global asymptotic stability of the system about its zero fixed point.

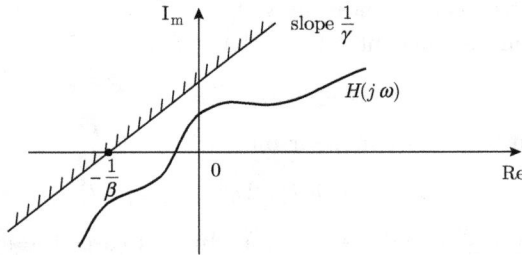

Fig. 4.5   Geometric meaning of the Popov criterion.

Example 4.2. Consider again the linear system discussed in Example 4.1. This 1-dimensional system is controllable as long as $b \neq 0$. It is also easy to verify that the system satisfies the global sector condition, even with $\alpha = 0$. Finally, for any $\varepsilon > 0$, if the constant γ is chosen such that

$$\gamma > \frac{\varepsilon\left(a^2 b^2 c^2 + \omega^2\right) - abc - a^2 b^2 c^2 - \omega^2}{\omega^2},$$

then some simple calculation shows that

$$\text{Re} \left\{ \frac{1 + j\,\gamma\,\omega}{j\,\omega - (a + bc)} \right\} + 1 > \varepsilon \qquad \text{for all} \quad \omega \geq 0.$$

Therefore, this linear feedback control system is globally asymptotically stable about its zero fixed point. This is consistent with the familiar linear analysis.

Example 4.3. Consider a Lur'e system with linearity described by the transfer function $G(s) = 1/[s(s + a)^2]$ and nonlinearity described by the relay function $h(\cdot)$ shown in Fig. 4.6, where a and $0 < b < 1$ are constants. Clearly,

$$G_r(j\,\omega) = \frac{-2a}{(a^2 + \omega^2)^2} \qquad \text{and} \qquad G_i(j\,\omega) = \frac{-\omega(a^2 - \omega^2)^2}{(a^2 + \omega^2)^2},$$

and the Popov criterion requires

$$\frac{1}{\beta} > \frac{2a}{(a^2 + \omega^2)^2} - \frac{\gamma\omega^2(a^2 - \omega^2)^2}{(a^2 + \omega^2)^2}$$

for all $\omega \geq 0$, which leads to the choice of a small $\gamma > 0$ with

$$\alpha = 0 \qquad \text{and} \qquad \beta = a^3/2.$$

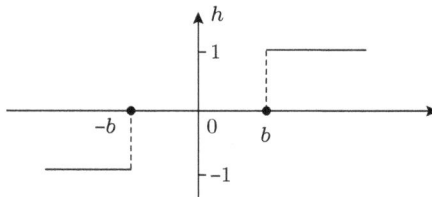

Fig. 4.6   The relay nonlinearity.

The Popov criterion has a natural connection to the linear Nyquist criterion. A more direct generalization of the Nyquist criterion to nonlinear systems is the following.

Theorem 4.2 (Circle Criterion). *Suppose that the SISO Lur'e system (4.9) satisfies the following conditions:*

(i) *matrix A has no purely imaginary eigenvalues, but has κ eigenvalues with positive real parts;*

(ii) *the system satisfies the global sector condition;*
(iii) *one of the following situations holds:*
  (a) $0 < \alpha < \beta$: *the Nyquist plot of* $G(j\,\omega)$ *encircles the disk* $D\left(-\frac{1}{\alpha}, -\frac{1}{\beta}\right)$ *counterclockwise* κ *times but does not enter it;*
  (b) $0 = \alpha < \beta$: *the Nyquist plot of* $G(j\,\omega)$ *stays within the open half-plane* $\mathrm{Re}\{s\} > -\frac{1}{\beta}$;
  (c) $\alpha < 0 < \beta$: *the Nyquist plot of* $G(j\,\omega)$ *stays within the open disk* $D\left(-\frac{1}{\beta}, -\frac{1}{\alpha}\right)$;
  (d) $\alpha < \beta < 0$: *the Nyquist plot of* $-G(j\,\omega)$ *encircles the disk* $D\left(\frac{1}{\alpha}, \frac{1}{\beta}\right)$ *counterclockwise* κ *times but does not enter it.*

Then, the system is globally asymptotically stable about its fixed point $\mathbf{x}^* = 0$.

Here, the disk $D\left(-\frac{1}{\alpha}, -\frac{1}{\beta}\right)$, for the case of $0 < \alpha < \beta$, can be visualized in Fig. 4.7.

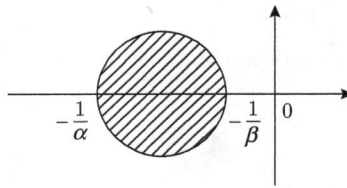

Fig. 4.7   The disk $D\left(-\frac{1}{\alpha}, -\frac{1}{\beta}\right)$.

Example 4.4. Consider the Lur'e system

$$\begin{cases} \dot{x} = a\,x + h(u) \\ y = b\,x + h(u) \\ u = k_1 x + k_2 y, \end{cases}$$

where $a < 0$, $b < 0$, k_1 and k_2 are constants, and the nonlinear function $h(\cdot)$ satisfies the global sector condition with $\alpha = 0$ and $\beta = \infty$.

It follows from a direct calculation that

$$\mathrm{Re}\left\{ (1 + j\,\gamma\omega)\,G(j\,\omega) \right\} + \frac{1}{\beta}$$
$$= \frac{(ak_2 + bk_2)\,(ab - \omega^2) + (k_1 + k_2)\omega^2(a + b)}{(ab - \omega^2)^2 + \omega^2(a + b)^2}.$$

Thus, the circle criterion (4.12) is

$$(ak_2 + bk_2)\, ab - \omega^2 \left[(ak_1 + bk_2) + (k_1 + k_2)\,(a + b) \right] > 0$$

for all $\omega \geq 0$, which is equivalent to

$$ak_2 + bk_2 > 0 \qquad \text{and} \qquad (ak_1 + bk_2) + (k_1 + k_2)\,(a + b) > 0 \,.$$

4.2.3   MIMO Lur'e Systems

Consider a multi-input multi-output (MIMO) Lur'e system, as shown in Fig. 4.3, of the form

$$\begin{cases} \mathbf{x}(s) = G(s)\,\mathbf{u}(s) \\ \mathbf{y} = C\,\mathbf{x} \\ \mathbf{u}(t) = -\,\mathbf{h}(\mathbf{y}(t)) \,, \end{cases} \tag{4.13}$$

where $G(s)$, as well as $\mathbf{h}$ and C, are defined in (4.2). If the system satisfies the following Popov inequality:

$$\int_{t_0}^{t_1} \mathbf{y}^{\top}(\tau)\mathbf{x}(\tau)\, d\tau \geq -\gamma \qquad \text{for all} \quad t_1 \geq t_0 \tag{4.14}$$

for a constant $\gamma \geq 0$ independent of t, then it is said to be *hyperstable*.

The linear part of this MIMO system is described by the transfer matrix $G(s)$, which is said to be *positive real* if

(i) $G(s)$ has no poles located inside the open half-plane $\text{Re}\{s\} > 0$;
(ii) poles of $G(s)$ on the imaginary axis are simple, and the residues of its entries (as $\omega \to \infty$) form a semi-positive definite and symmetric matrix;
(iii) the matrix $G_r(j\,\omega) := \frac{1}{2}\left[G(j\,\omega) + G^{\top}(-j\,\omega) \right]$ is a semi-positive definite and symmetric matrix for all real values of ω that are not poles of $G(s)$.

Example 4.5. The transfer matrix

$$G(s) = \begin{bmatrix} s/(s+1) & -1/(s+2) \\ 1/(s+2) & 2/(s+1) \end{bmatrix}$$

is positive real, because (i) all its poles are located on the left-half plane, (ii) the matrix formed by the residues,

$$G_r(j\infty) = \begin{bmatrix} 1 & 0 \\ 0 & 0 \end{bmatrix},$$

is semi-positive definite, and (iii) the matrix

$$G_r(j\,\omega) = \begin{bmatrix} \omega^2/(1+\omega^2) & j\omega/(4+\omega^2) \\ -j\,\omega/(4+\omega^2) & 2/(1+\omega^2) \end{bmatrix}$$

is positive definite for all $\omega \geq 0$ (its smallest eigenvalue approaches 2 as $\omega \to \infty$).

The matrix shown in this example is sometimes referred to as *strictly positive real*, which satisfies stronger conditions: (i)$'$ has no pole on the imaginary axis, (ii)$'$ the residues of its entries form a nonzero, semi-positive definite and symmetric matrix, and (iii)$'$ the matrix $G(j\,\omega) + G^\top(-j\,\omega)$ is positive definite and symmetric for all $\omega \geq 0$.

Theorem 4.3 (Hyperstability Theorem). *The MIMO Lur'e system (4.13) is hyperstable if and only if its transfer matrix $G(s)$ is positive real. Moreover, it is asymptotically hyperstable if and only if its transfer matrix is strictly positive real.*

Proof. See [Anderson and Vongpanitterd (1972)]; [Popov (1973)]. $\square$

4.3 Harmonic Balance Approximation and Describing Function

Consider again the SISO Lur'e systems (4.9), namely,

$$\begin{cases} \dot{\mathbf{x}} = A\,\mathbf{x} + \mathbf{b}\,u \\ y = \mathbf{c}^\top \mathbf{x} \\ u = h(y)\,, \end{cases} \tag{4.15}$$

where $h(0) = 0$, so $\mathbf{x}^* = 0$ is a fixed point of the system. Assume that the matrix A is invertible.

The interest here is the existence of periodic orbits (e.g. limit cycles) in the system output $y(t)$. Presumably, the system output has a periodic orbit, which is expressed in the Fourier series form:

$$y(t) = \sum_{k=-\infty}^{\infty} a_k\, e^{j\,k\,\omega t}\,,$$

where all $a_k = \bar{a}_k$ and ω are to be determined.

Observe that, if $y(t)$ is periodic, since $h(\cdot)$ is a time-invariant function, $h(y(t))$ is also periodic with the same period ω. Therefore,

$$u(t) = h(y(t)) = \sum_{k=-\infty}^{\infty} c_k\, e^{j\,k\,\omega t}\,,$$

where all $c_k = c_k(a_0, a_{\pm 1}, \cdots)$ are to be determined.

Observe also that in the frequency s-domain the input–output relation of the linear plant is

$$Y(s) = G(s)\,U(s)\,,$$

where the transfer matrix

$$G(s) = \mathbf{c}^\top \left[sI - A\right]^{-1} \mathbf{b} := \frac{n(s)}{d(s)}$$

is equivalent to a differential equation of constant coefficients with zero initial conditions:

$$d(p)\, y(t) = n(p)\, u(t)\,, \qquad p := d/dt\,.$$

Since

$$p\, e^{j\,k\,\omega t} = \frac{d}{dt}\, e^{j\,k\,\omega t} = j\,k\,\omega\, e^{j\,k\,\omega t}\,,$$

one has

$$d(p)\, y(t) = \sum_{k=-\infty}^{\infty} d(j\,k\,\omega)\, a_k\, e^{j\,k\,\omega t}$$

and

$$n(p)\, u(t) = \sum_{k=-\infty}^{\infty} n(j\,k\,\omega)\, c_k\, e^{j\,k\,\omega t}\,.$$

Consequently, the above differential equation yields

$$\sum_{k=-\infty}^{\infty} \left[d(j\,k\,\omega)\, a_k - n(j\,k\,\omega)\, c_k \right] e^{j\,k\,\omega t} = 0\,.$$

It then follows from the orthogonality of the Fourier basis functions that

$$d(j\,k\,\omega)\, a_k - n(j\,k\,\omega)\, c_k = 0\,, \qquad k = 0, \pm 1, \pm 2, \ldots\,,$$

or

$$G(j\,k\,\omega)\, c_k - a_k = 0\,, \qquad k = 0, \pm 1, \pm 2, \ldots\,.$$

Here, for $k = 0$, $G(0)$ is well defined since A is invertible by assumption.

Observe, furthermore, that since $G(j\,k\,\omega) = \overline{G}(-j\,k\,\omega)$, $a_k = \bar{a}_k$, and $c_k = \bar{c}_k$, for all $k = 0, \pm 1, \pm 2, \ldots$, it suffices to consider

$$G(j\,k\,\omega)\, c_k - a_k = 0\,, \qquad k = 0, 1, 2, \ldots\,.$$

However, this infinite-dimensional system of nonlinear algebraic equations is very different if not impossible to solve for the unknown ω and $\{a_k\}$, while $\{c_k\}$ are functions of $\{a_k\}$. One thus resorts to applying some kind of approximations.

Since $G(s)$ is strictly proper, it satisfies

$$\left| G(j\,k\,\omega) \right| \to 0 \qquad (k \to \infty)\,,$$

namely, $G(k\,j\,\omega) \approx 0$ for large values of k. So, it is reasonable to truncate the above infinite-dimensional system. The first-order approximation, with $k = 0$ and 1, gives

$$\begin{cases} G(0)\,\widehat{c}_0 - \widehat{a}_0 = 0 \\ G(j\,\omega)\,\widehat{c}_1 - \widehat{a}_1 = 0 \,, \end{cases}$$

where the first equation is real and the second is complex, with $\widehat{c}_i \approx c_i$ and $\widehat{a}_i \approx a_i$, $i = 0, 1$. Because $\widehat{c}_0$ and $\widehat{c}_1$ are both functions of $\widehat{a}_0$ and $\widehat{a}_1$, the above system of two equations has only two real unknowns, ω and $\widehat{a}_0$, and one complex unknown, $\widehat{a}_1$. Clearly, to solve for four real unknowns from three real equations, one more constraint is needed. It turns out that if one only considers the following special cases of the nonlinear feedback system, then some useful results can be obtained: Assume that

(i) the nonlinear function $h(\cdot)$ is odd: $h(-y) = -h(y)$, and is time-invariant;
(ii) if $y = \alpha \sin(\omega t)$, $\alpha \neq 0$, then the first-order harmonic of $-h(y)$ dominates the other harmonic components.

Here, assumption (i) implies that $\widehat{c}_0 = 0$, so that the first real equation yields $\widehat{a}_0 = 0$ as well. Assumption (ii) implies that

$$\alpha \sin(\omega t) = \frac{\alpha}{2j} \left(e^{j\,\omega t} - e^{-j\,\omega t} \right),$$

which gives $\widehat{a}_1 = \alpha/(2j)$. Thus, the second equation above becomes

$$G(j\,\omega)\,\widehat{c}_1(0, \alpha/(2j)) - \alpha/(2j) = 0 \,,$$

where, by the Fourier series coefficient formulas,

$$\begin{aligned} \widehat{c}_1(0, \alpha/(2j)) &= \frac{\omega}{2\pi} \int_0^{2\pi/\omega} -h(\alpha(\omega t))\, e^{-j\,\omega t} dt \\ &= j\,\frac{\omega}{\pi} \int_0^{\pi/\omega} h(\alpha(\omega t))\, \sin(\omega t)\, dt \,. \end{aligned}$$

Define

$$\Psi(\alpha) := \frac{\widehat{c}_1(0, \alpha/(2j))}{\alpha/(2j)} = -\frac{2\omega}{\pi\,\alpha} \int_0^{\pi/\omega} h(\alpha \sin(\omega t))\, \sin(\omega t)\, dt \,, \qquad (4.16)$$

which is called the *describing function* of the odd nonlinearity $-h(\cdot)$. Then, one obtains the first-order *harmonic balance equation*:

$$G(j\,\omega)\,\Psi(\alpha) + 1 = 0 \,. \qquad (4.17)$$

If this complex equation is solvable, then one may first solve

$$\begin{cases} G_r(j\,\omega)\,\Psi(\alpha) + 1 = 0 \\ G_i(j\,\omega) = 0\,, \end{cases}$$

for the two real unknowns ω and α, where G_r and G_i are the real and imaginary parts of G, respectively. This yields all the other expected results:

$$\widehat{a}_1 = \alpha/(2j)\,, \quad \widehat{c}_1 = j\,\frac{\omega}{\pi} \int_0^{\pi/\omega} h(\alpha(\omega t))\,\sin(\omega t)\,dt\,, \quad \widehat{a}_0 = 0\,, \quad \widehat{c}_0 = 0\,.$$

Consequently,

$$\widehat{y}(t) \approx \widehat{a}_{-1} e^{-j\,\omega t} + \widehat{a}_1 e^{j\,\omega t} = \frac{j\,\alpha}{2} e^{-j\,\omega t} - \frac{j\,\alpha}{2} e^{j\,\omega t}$$

is the first-order approximation of a *possible* periodic orbit of the system output, which is generated by an input signal of the form

$$\widehat{u}(t) = h(\widehat{y}(t)) \approx \widehat{c}_{-1} e^{-j\,\omega t} + \widehat{c}_1 e^{j\,\omega t}\,.$$

In summary, one arrives at the following conclusion, which is helpful for predicting possible periodic orbits (limit cycles) of the system output.

Theorem 4.4. *Consider the SISO Lur'e system*

$$\begin{cases} \dot{\mathbf{x}} = A\mathbf{x} + \mathbf{b}\,u \\ \mathbf{y} = \mathbf{c}^\top \mathbf{x} \\ u = h(\mathbf{y})\,. \end{cases}$$

Assume that the nonlinear function $h(\cdot)$ is odd and time-invariant, with the property that for $y(t) = \alpha \sin(\omega t)$, $\alpha \neq 0$, only the first-order harmonic of $-h(y)$ is significant. Define the describing function of $-h(\cdot)$ by

$$\Psi(\alpha) = -\frac{2\omega}{\pi\,\alpha} \int_0^{\pi/\omega} h(\alpha \sin(\omega t))\,\sin(\omega t)\,dt\,.$$

If the first-order harmonic balance equation

$$\begin{cases} G_r(j\,\omega)\,\Psi(\alpha) + 1 = 0 \\ G_i(j\,\omega) = 0 \end{cases}$$

has solutions ω and α, then

$$\widehat{y}^{\langle 1 \rangle}(t) = \frac{j\,\alpha}{2} e^{-j\,\omega t} - \frac{j\,\alpha}{2} e^{j\,\omega t}$$

is the first-order approximation of a possible periodic orbit of the system output. If this harmonic balance equation does not have solutions, then the system will not likely output any periodic orbits.

Example 4.6. Consider an SISO system with

$$G(s) = \frac{1}{s(s+1)(s+2)}$$

and

(a) as shown in Fig. 4.8(a):

$$-h(y) = \text{sgn}(y) = \begin{cases} 1 & y > 0 \\ 0 & y = 0 \\ -1 & y < 0 \,; \end{cases}$$

(b) as shown in Fig. 4.8(b):

$$-h(y) = \text{sat}(y) = \begin{cases} -1 & y < -1 \\ y & -1 \le y \le 1 \\ 1 & 1 < y \,. \end{cases}$$

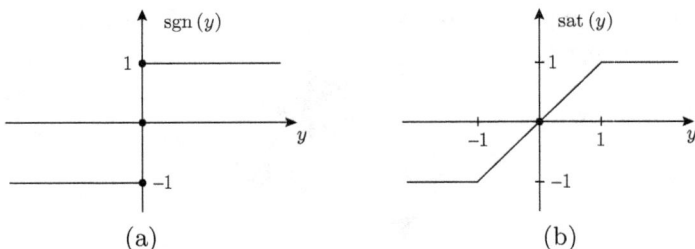

Fig. 4.8   Two different functions of $-h(y)$.

First, it can be verified that

$$G(j\,\omega) = \frac{1}{j\,\omega\,(j\,\omega+1)\,(j\,\omega+2)} = \frac{-3\omega - j\,(2-\omega^2)}{9\omega^2 + \omega(2-\omega^2)^2}\,,$$

so that the second harmonic balance equation becomes

$$G_i(j\,\omega) = \frac{-(2-\omega^2)}{9\omega^3 + \omega(2-\omega^2)^2} = 0\,,$$

which has two roots: $\omega_1 = \sqrt{2}$ and $\omega_2 = -\sqrt{2}$ (this one is ignored due to symmetry).

(a) The describing function in this case with a signum function is

$$\Psi(\alpha) = \frac{2}{\pi\,\alpha} \int_0^\pi -h(\alpha \sin\theta) \sin\theta \, d\theta = \frac{4}{\pi\,\alpha} \qquad (\theta = \omega t)\,.$$

Hence, the first harmonic balance equation at $\omega = \sqrt{2}$ becomes

$$\frac{-3\sqrt{2}}{9(\sqrt{2})^3 + \sqrt{2}\left(2 - (\sqrt{2})^2\right)^2} \frac{4}{\pi\alpha} + 1 = 0,$$

which yields $\alpha = 2/(3\pi)$. The conclusion is that there is possibly a periodic orbit of the form

$$\widehat{y}^{\langle 1 \rangle}(t) = \frac{j\alpha}{2} e^{-j\omega t} - \frac{j\alpha}{2} e^{j\omega t} = \frac{j}{3\pi} e^{-j\sqrt{2}t} - \frac{j}{3\pi} e^{j\sqrt{2}t},$$

which has amplitude $2/(3\pi)$ and frequency $\sqrt{2}$.

(b) The describing function in this case with a saturation function is

$$\Psi(\alpha) = \frac{2}{\pi\alpha} \int_0^\pi -h(\alpha\sin\theta)\sin\theta\, d\theta \le 1 \qquad \text{for all } \alpha,$$

so that the first harmonic balance equation at $\omega = \sqrt{2}$ becomes

$$\frac{-3\sqrt{2}}{9\left(\sqrt{2}\right)^3 + \sqrt{2}\left(2 - (\sqrt{2})^2\right)^2} \Psi(\alpha) + 1 = 0, \qquad \Psi(\alpha) \le 1,$$

which has no solution for all real (unknown) α. The conclusion is that there does not likely exist any periodic orbit in the system output. To make sure this is the case, usually higher-order harmonic balance approximations are needed, so as to obtain more accurate predictions (see [Moiola and Chen (1996)]).

When solving the equation $G_r(j\omega)\Psi(\alpha) = 1$ graphically, one can sketch two curves on the complex plane, $G_r(j\omega)$ and $-1/\Psi(\alpha)$, and then gradually increase ω and α respectively to find their crossing points:

(i) if the two curves are (visually) tangent, as illustrated in Fig. 4.9(a), then a conclusion drawn from the describing function method will not be satisfactory in general;

(ii) if the two curves are (visually) transversal, as illustrated in Fig. 4.9(b), then a conclusion drawn from the describing function analysis will generally be reliable.

Theorem 4.5 (Graphical Stability Criterion for Periodic Orbits).
Each intersection point of the two curves, $G_r(j\omega)$ and $-1/\Psi(\alpha)$, corresponds to a periodic orbit, $\widehat{y}^{\langle 1 \rangle}(t)$, in the output of system (4.15). If the points, near the intersection and on one side of the curve $-1/\Psi(\alpha)$ where $-\alpha$ is increasing, are not encircled by the curve $G_r(j\omega)$, then the corresponding periodic output is stable; otherwise, it is unstable.

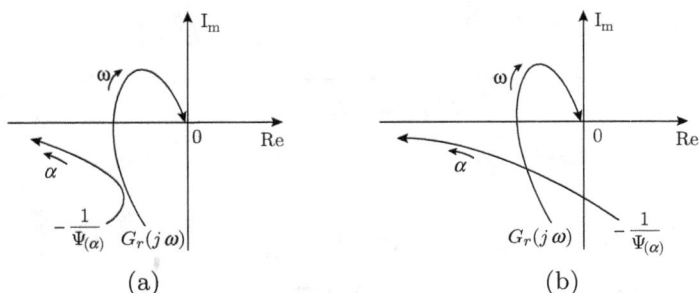

Fig. 4.9   Graphical describing function analysis.

Proof. See [Moiola and Chen (1996)]: pp. 25–26.   □

A word of warning is that the describing method discussed in this section is a graphical method, which is based on numerical computation and human visual judgement; therefore, although convenient and mostly successful in practice, there is a chance that it leads to incorrect conclusions about the system stability. A counterexample is given in Exercise 4.4.

4.4   BIBO Stability

A relatively simple and also relatively weak, but very practical type of stability is discussed in this section.

This is the *bounded-input bounded-output (BIBO) stability*, which refers to the property of a system that any bounded input to the system produces a bounded output through the system.

Consider an input–output map and its configuration as shown in Fig. 4.10(a).

Definition 1. The system S is said to be BIBO stable from the input set U to the output set Y if, for each admissible input $\mathbf{u} \in U$ and the corresponding output $\mathbf{y} \in Y$, there exist two non-negative constants, b_i and b_o, such that

$$||\mathbf{u}||_U \leq b_i \qquad \Longrightarrow \qquad ||\mathbf{y}||_Y \leq b_o. \qquad (4.18)$$

Note that since all norms are equivalent in a finite-dimensional vector space, it is unnecessary to distinguish under what kind of norms for the input and output signals the BIBO stability is defined and measured. Moreover, it is important to note that, in the above definition, even if b_i

is small and b_o is large, the system is still considered to be BIBO stable, which is usually good enough for many practical applications.

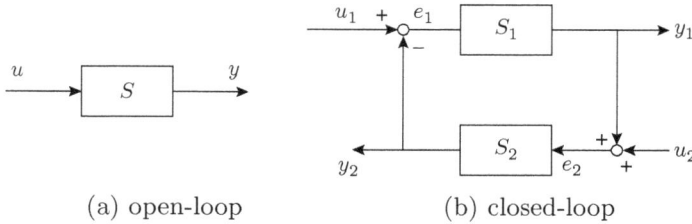

(a) open-loop (b) closed-loop

Fig. 4.10 Input–output relations.

4.4.1 *Small Gain Theorem*

A convenient criterion for verifying the BIBO stability of a closed-loop control system is the small gain theorem, which applies to most systems (linear and nonlinear, continuous-time and discrete-time, deterministic and stochastic, time-delayed, of any dimensions) in various forms, as long as the mathematical setting is appropriately formulated to meet the conditions of the theorem. The main disadvantage of this criterion is its over-conservativity, in general.

Consider the typical closed-loop system shown in Fig. 4.10(b), where the inputs, outputs, and internal signals are related as follows:

$$\begin{cases} S_1(e_1) = e_2 - u_2 \\ S_2(e_2) = u_1 - e_1 . \end{cases} \tag{4.19}$$

First, it is important to note that the individual BIBO stability of S_1 and S_2 is not sufficient to guarantee the BIBO stability of the connected closed-loop system. For instance, in the discrete-time setting of Fig. 4.10(b), suppose that $S_1 \equiv 1$ and $S_2 \equiv -1$, with $u_1(k) \equiv 1$ for all $k = 0, 1, \ldots$. Then, S_1 and S_2 are BIBO stable individually, but it can be easily verified that $y_1(k) = k \to \infty$ as the discrete-time variable k evolves. Therefore, a stronger condition restricting the interaction of S_1 and S_2 is necessary.

Theorem 4.6 (Small Gain Theorem). *If there exists four constants,* $L_1, L_2, M_1, M_2,$ *with* $L_1 L_2 < 1,$ *such that*

$$\begin{cases} ||S_1(e_1)|| \le M_1 + L_1||e_1|| \\ ||S_2(e_2)|| \le M_2 + L_2||e_2|| , \end{cases} \tag{4.20}$$

then

$$\begin{cases} ||e_1|| \le (1 - L_1 L_2)^{-1}\Big(||u_1|| + L_2||u_2|| + M_2 + L_2 M_1 \Big) \\ ||e_2|| \le (1 - L_1 L_2)^{-1}\Big(||u_2|| + L_1||u_1|| + M_1 + L_1 M_2 \Big), \end{cases} \tag{4.21}$$

where the norms $||\cdot||$ *are defined over the spaces to which the signals belong. Consequently, (4.20) and (4.21) together imply that if the system inputs (u_1 and u_2) are bounded then the corresponding outputs ($S_1(e_1)$ and $S_2(e_2)$) are bounded.*

Note that the four constants, L_1, L_2, M_1, M_2, can be somewhat arbitrary (e.g. either L_1 or L_2 can be large, and some of them can even be negative). As long as $L_1 L_2 < 1$, the BIBO stability conclusion follows. This inequality is the key condition for the theorem to hold, which is required by the inversion $(1 - L_1 L_2)^{-1}$ in the bounds (4.21).

Proof. It follows from (4.19) and (4.20) that

$$||e_1|| \le ||u_1|| + ||S_2(e_2)|| \le ||u_1|| + M_2 + L_2||e_2||.$$

Similarly, one has

$$||e_2|| \le ||u_2|| + ||S_1(e_1)|| \le ||u_2|| + M_1 + L_1||e_1||.$$

By combining these two inequalities, one obtains

$$||e_1|| \le L_1 L_2||e_1|| + ||u_1|| + L_2||u_2|| + M_2 + L_2 M_1,$$

which, on the basis of $L_1 L_2 < 1$, yields

$$||e_1|| \le (1 - L_1 L_2)^{-1}\big(||u_1|| + L_2||u_2|| + M_2 + L_2 M_1 \big).$$

The other inequality can be similarly verified. $\qquad\qquad\square$

In the special case where the input–output spaces, U and Y, are both in the L_2-space, a similar criterion based on the system passivity property can be obtained. In this case, an inner product between two vectors in the space will be useful, which is defined by

$$\langle \xi, \eta \rangle = \int_{t_0}^{\infty} \xi^{\top}(\tau)\eta(\tau)d\tau.$$

Theorem 4.7 (Passivity Stability Theorem). *If there are four constants, L_1, L_2, M_1, M_2, with $L_1 + L_2 > 0$, such that*

$$\begin{cases} \langle\, e_1, S_1(e_1) \,\rangle \ge L_1||e_1||^2 + M_1 \\ \langle\, e_2, S_2(e_2) \,\rangle \ge L_2||S_2(e_2)||^2 + M_2, \end{cases} \tag{4.22}$$

then the closed-loop system (4.19) is BIBO stable.

Proof. See [Sastry (1999)]: pp. 155–156. □

As mentioned above, the main disadvantage of this criterion is its over-conservativeness in providing the sufficient conditions for the BIBO stability. One resolution is to transform the system into the Lur'e configuration, and then apply the circle or Popov criterion under the sector condition, if it can be satisfied, which may lead to less-conservative stability conditions, in general.

4.4.2 Relation between BIBO and Lyapunov Stabilities

There is a closed relationship between the BIBO stability of a nonlinear feedback system and the Lyapunov stability of a related nonlinear control system [Desoer and Vidyasagar (1975)].

Consider a nonlinear system in the form of

$$\dot{\mathbf{x}} = A\,\mathbf{x} - \mathbf{f}(\mathbf{x}, t)\,, \qquad \mathbf{x}(t_0) = \mathbf{x}_0\,, \tag{4.23}$$

where $\mathbf{f} : R^n \times R^1 \to R^n$ is a real vector-valued integrable nonlinear function, satisfying $\mathbf{f}(0, t) = 0$ for all $t \in [0, \infty)$ (so the system has a zero fixed point). Assume also that the system matrix A is stable (i.e. has all eigenvalues with negative real parts).

By adding and then subtracting the term $A\mathbf{x}$, a general nonlinear system can always be written in this form.

Now, define

$$\begin{cases} \mathbf{x}(t) = \mathbf{u}(t) - \int_{t_0}^{t} e^{(t-\tau)A}\,\mathbf{y}(\tau)d\tau \\ \mathbf{y}(t) = \mathbf{f}(\mathbf{x}, t)\,, \end{cases} \tag{4.24}$$

with $\mathbf{u}(t) = \mathbf{x}_0 e^{tA}$. Then, system (4.23) can be implemented by a feedback configuration as depicted in Fig. 4.11, with error signal $\mathbf{e} = \mathbf{x}$, plant $P(\cdot)(t) = \mathbf{f}(\cdot, t)$, and a compensator $C(\cdot)(t) = \int_{t_0}^{t} e^{(t-\tau)A}(\cdot)(\tau)d\tau$.

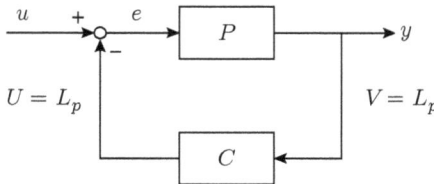

Fig. 4.11   A nonlinear feedback system.

Consider the nonlinear system (4.24) with the feedback configuration shown in Fig. 4.11.

Theorem 4.8. *Suppose that $U = V = L_p([0, \infty), R^n)$, with $1 < p < \infty$. Then, if the feedback system shown in Fig. 4.11 is BIBO stable from U to V, the nonlinear system (4.23) is globally asymptotically stable above the zero fixed point, i.e. $\|\mathbf{x}(t)\| \to 0$ as $t \to \infty$.*

Proof. Since all eigenvalues of the constant matrix A have negative real parts, one has

$$\| \mathbf{x}_0 e^{tA} \| \leq M\, e^{-\alpha t}$$

for some $0 < \alpha, M < \infty$ for all $t \in [0, \infty)$, so that $\|\mathbf{u}(t)\| = \|\mathbf{x}_0 e^{tA}\| \to 0$ as $t \to \infty$. Hence, in view of the first equation of (4.24), if it can be proven that

$$\mathbf{v}(t) := \int_0^t e^{(t-\tau)A}\, \mathbf{y}(\tau)\, d\tau \to 0 \qquad (t \to \infty)$$

in the Euclidean norm, then it will follow that

$$\|\mathbf{x}(t)\| = \|\mathbf{u}(t) - \mathbf{v}(t)\| \to 0 \qquad (t \to \infty)\,.$$

To prove so, write

$$\mathbf{v}(t) = \int_0^{t/2} e^{(t-\tau)A}\mathbf{y}(\tau)d\tau + \int_{t/2}^t e^{(t-\tau)A}\, \mathbf{y}(\tau)\, d\tau$$

$$= \int_{t/2}^t e^{\tau A}\, \mathbf{y}(t - \tau)\, d\tau + \int_{t/2}^t e^{(t-\tau)A}\, \mathbf{y}(\tau)\, d\tau\,.$$

Then, by the Hölder inequality, one has

$$\|\mathbf{v}(t)\| \leq \left\| \int_{t/2}^t e^{\tau A}\mathbf{y}(t-\tau)d\tau \right\| + \left\| \int_{t/2}^t e^{(t-\tau)A}\mathbf{y}(\tau)d\tau \right\|$$

$$\leq \left(\int_{t/2}^t |e^{\tau A}|^q\, d\tau \right)^{1/q} \left(\int_{t/2}^t |\mathbf{y}(t-\tau)|^p\, d\tau \right)^{1/p}$$

$$+ \left(\int_{t/2}^t \|e^{(t-\tau)A}\|^q\, d\tau \right)^{1/q} \left(\int_{t/2}^t |y(\tau)|^p\, d\tau \right)^{1/p}$$

$$\leq \left(\int_{t/2}^\infty \|e^{\tau A}|^q\, d\tau \right)^{1/q} \left(\int_0^\infty |y(\tau)|^p\, d\tau \right)^{1/p}$$

$$+ \left(\int_0^\infty \|e^{\tau A}\|^q\, d\tau \right)^{1/q} \left(\int_{t/2}^\infty |\mathbf{y}(\tau)|^p\, d\tau \right)^{1/p}\,.$$

Since all eigenvalues of A have negative real parts and since the feedback system is BIBO stable from U to V, so that $\mathbf{y} \in V = L_p([0, \infty), R^n)$, one has

$$\lim_{t \to \infty} \int_{t/2}^{\infty} |e^{\tau A}|^q \, d\tau = 0$$

and

$$\lim_{t \to \infty} \int_{t/2}^{\infty} |\mathbf{y}(\tau)|^p \, d\tau = 0 \,.$$

Consequently, it follows that $||\mathbf{v}(t)|| \to 0$ as $t \to \infty$. $\square$

4.4.3 *Contraction Mapping Theorem*

The small gain theorem discussed above by nature is a kind of contraction mapping theorem. The contraction mapping theorem can be used to determine the BIBO stability property of a system described by a map in various forms, provided that the system (or the map) is appropriately formulated. The following is a typical (global) contraction mapping theorem.

Define the operator norm of the input–output map S by

$$||S|| := \sup_{\mathbf{x}_1 \neq \mathbf{x}_2} \frac{||S(\mathbf{x}_2) - S(\mathbf{x}_1)||}{||\mathbf{x}_2 - \mathbf{x}_1||} \,.$$

Theorem 4.9 (Contraction Mapping Theorem). *If the operator norm of the input–output map S satisfies $||S|| < 1$, then the system equation*

$$\mathbf{y}(t) = S(\mathbf{y}(t)) + \mathbf{c}$$

has a unique solution for any constant vector $\mathbf{c} \in R^n$. This solution satisfies

$$||\mathbf{y}||_\infty \leq \left(1 - ||S|| \right)^{-1} ||\mathbf{c}|| \,.$$

Moreover, the sequence of the iterations

$$\mathbf{y}_{k+1} = S(\mathbf{y}_k) \,, \qquad \mathbf{y}_0 \in R^n \,, \qquad k = 0, 1, \ldots \,,$$

satisfies

$$||\mathbf{y}_k|| \to 0 \qquad \text{as} \quad k \to \infty \,.$$

Proof. In the continuous-time setting, it is clear that

$$||\mathbf{y}|| \leq ||S(\mathbf{y})|| + ||\mathbf{c}|| \leq ||S|| \cdot ||\mathbf{y}|| + ||\mathbf{c}|| \,,$$

so that

$$\left(1 - ||S|| \right) ||\mathbf{y}|| \leq ||\mathbf{c}|| \,,$$

Since $||S|| < 1$, one has

$$||\mathbf{y}|| \leq \left(1 - ||S|| \right)||\mathbf{c}|| \qquad \text{for all} \ \ t \geq 0 \, .$$

Therefore, taking the supremum over $t \in [0, \infty)$ on both sides yields the expected result immediately.

In the discrete-time case, let $\mathbf{y}_0 \in R^n$ be arbitrarily given, so $||\mathbf{y}_0|| < \infty$. By iterations, since $||S|| < 1$, one obtains

$$||\mathbf{y}_k|| = ||S(\mathbf{y}_{k-1})|| = \cdots = ||S^k(\mathbf{y}_0)|| \leq ||S||^k \, ||\mathbf{y}_0|| \to 0 \qquad (k \to \infty) \, .$$

$\square$

Exercises

4.1 The following is a model of a perfectly mixed chemical reactor with a cooling coil, in which two first-order, consecutive, irreversible, and exothermic reactions $A \to B \to C$ occur:

$$\begin{cases} \dot{x} = -x + a(1-x)\,e^z \\ \dot{y} = -y + a(1-x)\,e^z - acy\,e^z \\ \dot{z} = -(1+\beta)\,z + ab(1-x)\,e^z + abc\alpha\,y\,e^z\,, \end{cases}$$

where all coefficients are constants with certain physical meanings. Reformulate this model in the Lur'e form, namely, assuming $\mathbf{b} = 0$ and $\mathbf{c} = 1$, find A, $h(\cdot)$, and $G(s)$.

4.2 Consider the following autonomous system:

$$\ddot{x} + 2\,\dot{x} + f(x) = 0\,,$$

where $f(\cdot)$ is a nonlinear differentiable function. Find the bounds a and b for $ax \le f(x) \le bx$, such that the zero fixed point of the system is asymptotically stable.

4.3 Consider the following system:

$$\begin{cases} \dot{x}_1 = x_2 \\ \dot{x}_2 = -2\,x_2 + h(y) \\ y = x_1\,, \end{cases}$$

where $h(\cdot)$ is a nonlinear function belonging to a sector $[0, \kappa]$ for some $\kappa > 0$. Reformulate this system into the Lur'e form and then use either Popov or circle criterion to discuss its absolute stability.

4.4 Verify that the following SISO Lur'e system is a counterexample to both the Aizerman–Kalman conjecture and the describing function method:

$$A = \begin{bmatrix} 0 & 1 \\ -1 & -1 \end{bmatrix}, \qquad \mathbf{b} = \begin{bmatrix} 0 \\ -1 \end{bmatrix}, \qquad \mathbf{c}^\top = [-1 \;\; -1]\,,$$

and

$$h(y) = \begin{cases} \left(1 - \dfrac{e^{-2}}{1+e^{-1}}\right)y & |y| < 1 \\[3mm] \left(1 - \dfrac{e^{-2|y|}}{|y|\,(1+e^{-|y|})}\right)y & |y| \ge 1\,. \end{cases}$$

[Hint: see [Narendra and Taylor (1973)]: pp. 69–72.]

4.5 Show that the transfer function
$$G(s) = \frac{w_n^2(s+a)}{s^2 + 2\zeta w_n s + w_n^2},$$
where $w_n > 0$ and $0 < \zeta < 1$, is strictly positive real if and only if $0 < a < 2\zeta w_n$.

4.6 Use the circle criterion to discuss the absolute stability of the following system:
$$\begin{cases} \dot{x} = -x - h(x+y) \\ \dot{y} = x - y - 2h(x+y), \end{cases}$$
where $h(\cdot)$ is a nonlinear differentiable function given by
$$-h(y) = \begin{cases} -a & y < -b < 0 \\ 0 & -b \le y \le b \\ a & b < y. \end{cases}$$

4.7 Consider the following autonomous nonlinear system:
$$\begin{cases} \dot{x}_1 = -a\,x_1 + x_2 - h(x_1) \\ \dot{x}_2 = -x_1 + x_3 \\ \dot{x}_3 = -a\,x_1 + b\,h(x_1) \\ y = x_1, \end{cases}$$
where $a > 0$, $b > 0$, and $h(\cdot)$ is a piecewise continuous function satisfying $h(0) = 0$ and $0 < h(z) < \kappa z$. Reformulate this system into the Lur'e form and then apply the Popov or circle criterion to discuss the absolute stability of the system. Is there any condition on the sector upper bound κ? [Hint: Discuss three cases: (a) $b < a^2$, (b) $b = a^2$, and (c) $b > a^2$, respectively.]

4.8 Find the describing function for each of the following odd nonlinear functions:
$$h(y) = y^5, \qquad h(y) = y^3|y|, \qquad h(y) = \sin(y).$$

4.9 Consider the piecewise-linear function shown in Fig. 4.12, which contains the two nonlinear functions of Fig. 4.8 as special cases. Verify that

(a) when $|\alpha| \le \delta$, the describing function $\Psi(\alpha) = s_1$;
(b) when $\alpha > \delta$,
$$\Psi(\alpha) = \frac{2(s_1 - s_2)}{\pi}\left[\sin^{-1}(\delta/\alpha) + \delta\alpha^{-1}\sqrt{1 - (\delta/\alpha)^2}\right] + s_2;$$

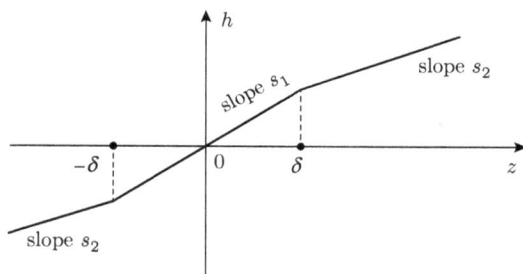

Fig. 4.12 A piecewise-linear function, $-h(y)$.

(c) with $s_1 = 1$, $s_2 = 0$, and $\delta = 1$, the above result reduces to that obtained in Example 4.6 for case (b).

4.10 Consider the nonlinear feedback system shown in Fig. 4.10(b), with

$$S_1 : e_1 \to \frac{e^{-t}}{2} \sin(e_1) \qquad \text{and} \qquad S_2 : e_2 \to \frac{e^{-t}}{2} + \sin(e_2).$$

Determine the upper bounds for $\|e_1\|$ and $\|e_2\|$, respectively, using the small gain theorem.

4.11 Consider a unity feedback system, as shown in Fig. 4.10(b), with $S_1 = P$ representing a given nonlinear plant and $S_2 = I$ (the identity mapping). Suppose that the plant has a perturbation, so P becomes $P + \Delta$. Derive the small gain theorem corresponding to this perturbed closed-loop system and find the norm condition on the perturbation term, namely, find the smallest possible upper bound on $\|\Delta\|$ in terms of the bounds given in the small gain theorem just derived, such that under this condition the perturbed system remains BIBO stable.

Chapter 5

Nonlinear Dynamics: Bifurcations and Chaos

Nonlinear systems have complex dynamics, especially bifurcations and chaos, in addition to periodic oscillations such as limit cycles, as briefly introduced in Sec. 1.4.

This chapter further studies some typical nonlinear dynamics, particularly bifurcations and chaos. Basic notions of period-doubling bifurcation and Hopf bifurcation are investigated, and strange attractors and chaos are discussed. Some typical chaotic systems are analyzed, mainly on continuous-time models but also briefly in the discrete-time setting. Throughout the course, some useful mathematical analysis tools such as normal form and Poincaré map are introduced.

5.1 Typical Bifurcations

Consider a nonlinear autonomous system with a real variable parameter:

$$\dot{\mathbf{x}} = \mathbf{f}(\mathbf{x}; \mu), \qquad \mathbf{x} \in R^n, \qquad \mu \in R, \qquad (5.1)$$

where the nonlinear function $\mathbf{f} \colon R^n \times R \to R^n$ is assumed to satisfy all necessary conditions for the existence (and oftentimes also the uniqueness) of a solution with respect to any given initial state $\mathbf{x}_0 \in R^n$ and fixed parameter $\mu \in R$. Assume also that $\mathbf{x}^* = 0$ is a fixed point, also called an *equilibrium point*, of the system at $\mu = \mu_0$, namely

$$\mathbf{f}(0, \mu_0) = 0.$$

The concern here is whether or not and, if so, how this equilibrium point and its stability may change as the parameter μ is gradually varied in a neighborhood of μ_0. This study has a great impact on the understanding of the dynamical behaviors and stabilities of the parameterized nonlinear system (5.1).

If there is a change of stability of the equilibrium point, as μ is varied and passes through a critical value μ_0, for instance the system is stable about its equilibrium point of interest when $\mu > \mu_0$, but becomes unstable when $\mu < \mu_0$, then the system is said to have a *bifurcation* at the equilibrium point. The critical value $\mu = \mu_0$ is called a *bifurcation value* and $(0, \mu_0)$ in the $\mathbf{x}$–μ space is called a *bifurcation point*. A more precise definition for the 1-dimensional case, namely

$$\dot{x} = f(x; \mu), \qquad x, \mu \in R, \tag{5.2}$$

is given below.

Definition 5.1. The one-parameter family of 1-dimensional nonlinear systems (5.2) is said to have a *bifurcation at an equilibrium point*, (x^*, μ_0), if the equilibrium curve near the point $x = x^*$ and near $\mu = \mu_0$ is not qualitatively the same as (i.e. not topologically equivalent to) the original curve near $x = x^*$ at $\mu = \mu_0$.

For illustration, a few simple and representative examples are discussed.

Example 5.1. Consider the 1-dimensional system

$$\dot{x} = f(x; \mu) = \mu\, x - x^2,$$

which has two equilibrium points: $x_1^* = 0$ and $x_2^* = \mu$. As μ is varied, two equilibrium curves will be generated, as shown in Fig. 5.1.

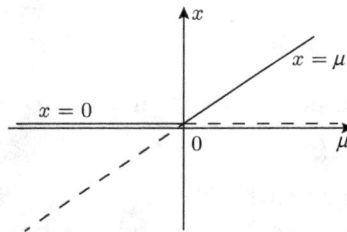

Fig. 5.1   Equilibrium curves of the system in Example 5.1.

The stabilities of the above equilibrium point and curves are determined as follows: first, since this is an autonomous system, one may calculate its Jacobian, obtaining

$$J = \left.\frac{\partial f}{\partial x}\right|_{x=x^*} = \mu - 2\,x\,\big|_{x=x^*}.$$

(i) At $x^* = 0$, $J = \mu$, so when $\mu < 0$, the system is stable about this equilibrium point and when $\mu > 0$, it is unstable.

(ii) At $x^* = \mu$, $J = -\mu$, so when $\mu < 0$, the system is unstable about this equilibrium point and when $\mu > 0$, it is stable.

Therefore, $(x^*, \mu_0) = (0, 0)$ is a bifurcation point. This type of bifurcation is called the *transcritical bifurcation*.

Example 5.2. Consider the 1-dimensional system
$$\dot{x} = f(x; \mu) = \mu - x^2 ,$$
which has one equilibrium point $x_0^* = 0$ at $\mu = \mu_0 = 0$, with an equilibrium curve $(x^*)^2 = \mu$ for all $\mu \geq 0$, which yields two branches $x_1^* = \sqrt{\mu}$ and $x_2^* = -\sqrt{\mu}$. Clearly, these two branches contain the point $x_0^* = 0$ when $\mu = 0$. As μ is varied over the interval $(0, \infty)$, these two equilibrium curves have shapes as shown in Fig. 5.2.

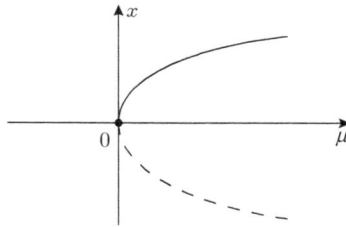

Fig. 5.2   Equilibrium curves of the system in Example 5.2.

The stabilities of the equilibrium curves can be similarly determined, as follows. The system Jacobian is
$$J = \frac{\partial f}{\partial x}\bigg|_{x=x^*} = -2x \big|_{x=x^*} .$$

(i) At $x^* = \sqrt{\mu}$, $J = -2\sqrt{\mu}$, so when $\mu > 0$, the system is stable about this equilibrium state.

(ii) At $x^* = -\sqrt{\mu}$, $J = 2\sqrt{\mu}$, so when $\mu > 0$, the system is unstable about this equilibrium state.

Therefore, $(x^*, \mu_0) = (0, 0)$ is a bifurcation point. This type of bifurcation is called the *saddle-node bifurcation*.

Example 5.3. Consider the 1-dimensional system
$$\dot{x} = f(x; \mu) = \mu x - x^3 ,$$

which has one equilibrium point $x_0^* = 0$ for all $\mu \in R$, with an equilibrium curve $(x^*)^2 = \mu$ for all $\mu \geq 0$, which yields two branches $x_1^* = \sqrt{\mu}$ and $x_2^* = -\sqrt{\mu}$. These two branches contain the point $x_0^* = 0$ when $\mu = 0$. As μ is varied, the two equilibrium curves have the shapes as shown in Fig. 5.3.

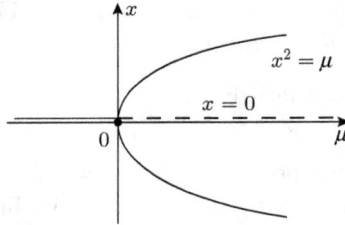

Fig. 5.3   Equilibrium curves of the system in Example 5.3.

The stabilities of the equilibrium curves can be similarly determined, as follows. The system Jacobian is

$$J = \left.\frac{\partial f}{\partial x}\right|_{x=x^*} = \left.\mu - 3x^2\right|_{x=x^*}.$$

(i)   At $x_0^* = 0$, $J = \mu$, so when $\mu > 0$, the system is unstable about this equilibrium state, while for $\mu < 0$, it is stable.

(ii) At $(x^*)^2 = \mu$, $J = -2\mu$, so when $\mu > 0$, the system is stable about both the two equilibrium branches $x_1^* = \sqrt{\mu}$ and $x_2^* = -\sqrt{\mu}$.

Therefore, $(x^*, \mu_0) = (0, 0)$ is a bifurcation point. This type of bifurcation is called the *pitchfork bifurcation*, inspired by the shape of the bifurcation diagram, as shown in Fig. 5.3.

To this end, it is important to note that not every nonlinear system with a varying parameter has bifurcations.

Example 5.4. Consider the 1-dimensional system

$$\dot{x} = f(x; \mu) = \mu - x^3,$$

which has one and only one equilibrium curve $(x^*)^3 = \mu$, for all $\mu \in R$. As μ is varied, this equilibrium curve has the shape as visualized in Fig. 5.4.

The stabilities of the equilibrium curve can be determined by examining the system Jacobian

$$J = \left.\frac{\partial f}{\partial x}\right|_{x=x^*} = \left.-3x^2\right|_{x=x^*} < 0 \qquad \text{for all} \quad x^* = \sqrt[3]{\mu} \neq 0.$$

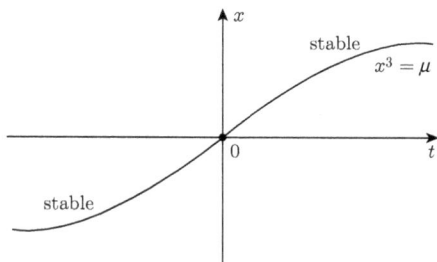

Fig. 5.4   The equilibrium curve of the system in Example 5.4.

Clearly, $(x^*)^3 = \mu$ is always stable and, around the critical point $(x^*, \mu_0) = (0,0)$, the stability of the system about the equilibrium curve does not change. Therefore, there is no bifurcation in this system.

It can be observed from the above typical bifurcation examples that (x^*, μ_0) is a bifurcation point for system (5.2) if either

(i) there are more than one curve of equilibrium solutions of the system passing through the point (x^*, μ_0) on the x–μ plane, or
(ii) in the case where there is only one equilibrium curve, this curve is located on one side of the vertical line $\mu = \mu_0$ in the neighborhood of the point (x^*, μ_0) on the x–μ plane.

For 1-dimensional one-parametric systems, this observation is generally correct.

5.2   Period-Doubling Bifurcation

For discrete-time systems, there is a special and interesting dynamical phenomenon called *period-doubling bifurcation*.

A typical example is the logistic map

$$x_{k+1} = \mu\, x_k(1 - x_k), \qquad k = 0, 1, 2, \ldots, \tag{5.3}$$

with $x_0 \in (0, 1)$. Let the non-negative parameter μ be gradually varied, starting from $\mu = 0$. Then, one can observe the following phenomena:

Case 1. $0 < \mu < 1$.

In this case, starting from any $x_0 \in (0, 1)$, one can observe that $x_k \to 0$ as $k \to \infty$.

Case 2. $1 \leq \mu < 3$.

Within this range of parameter values, starting from any $x_0 \in (0,1)$, one can observe that x_k approaches a steady state as $k \to \infty$. For instance, with $\mu = 2.6$, the result is shown in Fig. 5.5.

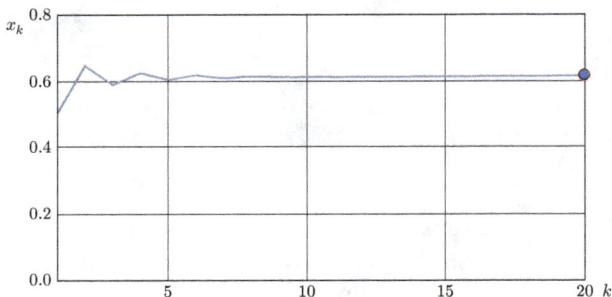

Fig. 5.5   Converging orbit of the logistic map with $\mu = 2.6$.

Case 3. $3 \leq \mu < 3.449 \cdots$.

Starting from any $x_0 \in (0,1)$, $x_k \to$ period-2 cycles as $k \to \infty$; namely, after a transient, $\{x_k\} = \{\cdots, x^{(1)}, x^{(2)}, x^{(1)}, x^{(2)}, x^{(1)}, x^{(2)}, \cdots\}$, as shown in Fig. 5.6 for the case of $\mu = 3.3$.

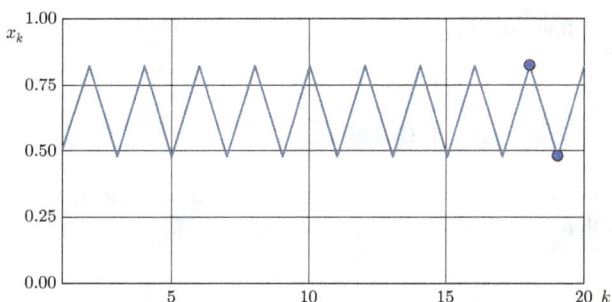

Fig. 5.6   Period-2 orbit of the logistic map with $\mu = 3.3$.

Case 4. $3.449 \cdots \leq \mu < 4.0$.

Depending on the value of μ, $x_k \to$ period-2^n cycle for some $n > 0$ as $k \to \infty$, as shown in Fig. 5.7 for the case of $\mu = 3.5$.

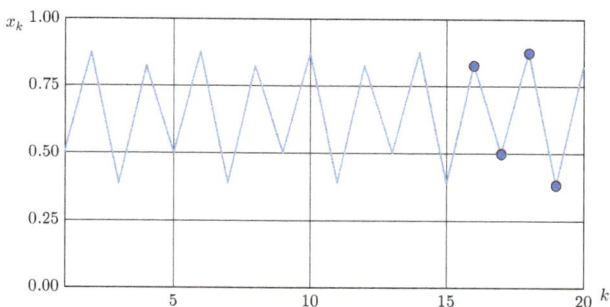

Fig. 5.7 Period-2^2 orbit of the logistic map with $\mu = 3.5$.

Within the range $(3, 4)$ of the parameter μ values, some very rich bifurcating phenomena, called the *period-doubling bifurcation*, can be observed.
The period-doubling bifurcated orbit is summarized in Table 5.1 and shown
in Fig. 5.8, and particularly by the μ–x_k diagram in Fig. 5.9, where very
interesting self-similarity within the period-doubling bifurcation is promi-
nent.

Table 5.1 Period-doubling bifurcation of the logistic map.

Parameter	Period	
$\mu < 3.0$	1	2^0
$\mu = 3.0$	2	2^1
$\mu = 3.449 \cdots$	4	2^2
$\mu = 3.54409 \cdots$	8	2^3
$\mu = 3.5644 \cdots$	16	2^4
$\mu = 3.568759 \cdots$	32	2^5
$\vdots$	$\vdots$	$\vdots$
$\mu = 3.569946 \cdots$	∞	2^∞

5.3 Hopf Bifurcations in 2-Dimensional Systems

Bifurcations in 2-dimensional parameterized systems can be quite complicated. Some examples of the hyperbolic systems are first studied, along
with the useful normal form theorem. For the non-hyperbolic case, the
system has a very typical Hopf bifurcation, which will be subsequently discussed.

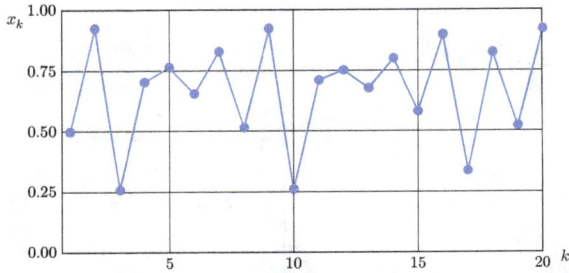

Fig. 5.8   Period-doubling of the logistic map with $\mu = 3.569946\cdots$.

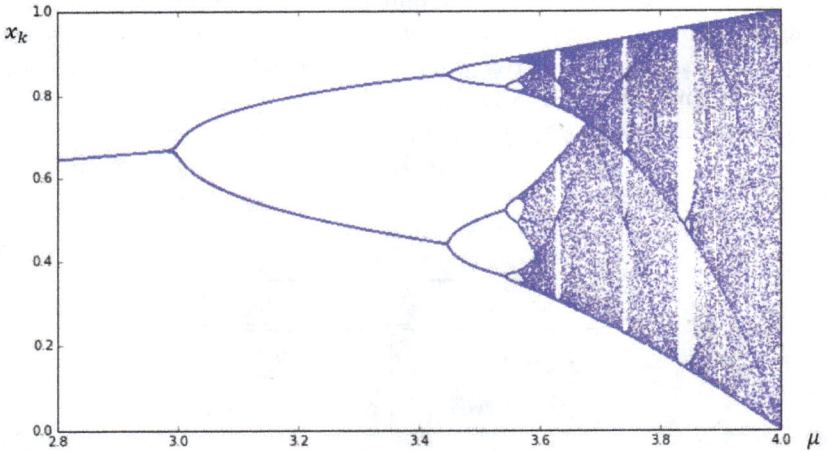

Fig. 5.9   Self-similarity of the logistic period-doubling diagram.

5.3.1   *Hyperbolic Systems and Normal Form Theorem*

Some examples of hyperbolic systems with rather complicated bifurcation phenomena are first presented and discussed.

Example 5.5. Consider the 2-dimensional linear parameterized system

$$\begin{cases} \dot{x} = -\mu x + y \\ \dot{y} = -\mu x - 3y, \end{cases} \qquad \mu \in R.$$

The system has eigenvalues

$$\lambda_{1,2} = -\frac{\mu + 3}{2} \pm \frac{1}{2}\sqrt{(\mu - 1)(\mu - 9)}\,.$$

Hence, the zero equilibrium point is hyperbolic. As μ is varied, some bifurcations occur as depicted in Fig. 5.10.

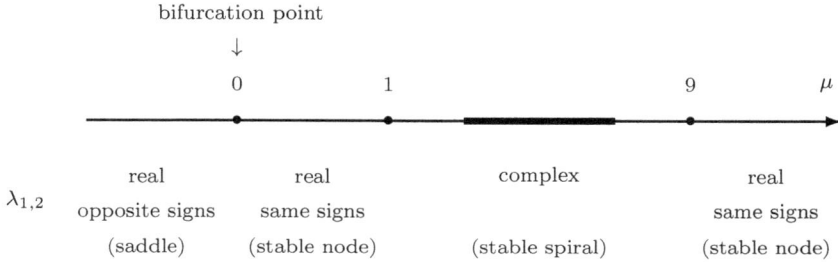

Fig. 5.10   Bifurcations of the 2-dimensional parametric system in Example 5.5.

Example 5.6. Consider the controlled damped pendulum

$$\begin{cases} \dot{x} = y \\ \dot{y} = -y - \sin(x) + u\,, \end{cases}$$

where the angular variable $x = \theta$ satisfies $-\pi \le \theta \le \pi$ and $y = \dot{\theta}$. This system has two possible equilibrium points: $y^* = 0$ and $\sin(x^*) = u$. There are three cases:

 (i) $u > 1$: there is no equilibrium point;
 (ii) $u = 1$: there is one equilibrium point, at $(x^*, y^*) = (\pi/2, 0)$;
 (iii) $u < 1$: there are two equilibrium points.

If $u = \mu$ is considered as the system variable parameter, then the above three cases together show that $\mu_0 = u_0 = 1$ is a bifurcation value.

In order to determine the type of bifurcation at μ_0, an effective way is to transform the system to the so-called *normal form*, in three steps, as follows:

Step 1. Shift (x^*, y^*, μ_0) to $(0, 0, 0)$

This can be done by a change of variables:

$$\begin{cases} u = 1 + v \\ x = z_1 + \pi/2 \\ y = z_2\,, \end{cases}$$

which yields a new system of the form

$$\begin{cases} \dot{z}_1 = z_2 \\ \dot{z}_2 = -z_1 - \cos(z_1) + 1 + v. \end{cases}$$

Step 2. Find the Jacobian and its eigenvalues and eigenvectors

$$J = \begin{bmatrix} 0 & 1 \\ \sin(z_1) & -1 \end{bmatrix}_{z_1^* = 0} = \begin{bmatrix} 0 & 1 \\ 0 & -1 \end{bmatrix}$$

and

$$\lambda_1 = 0, \quad \mathbf{w}_1 = \begin{bmatrix} 1 \\ 0 \end{bmatrix}; \qquad \lambda_2 = -1, \quad \mathbf{w}_2 = \begin{bmatrix} -1 \\ 1 \end{bmatrix}.$$

Step 3. Find the normal form

Use the nonsingular matrix

$$P := \begin{bmatrix} \mathbf{w}_1 & \mathbf{w}_2 \end{bmatrix} = \begin{bmatrix} 1 & -1 \\ 0 & 1 \end{bmatrix}$$

to transform the new system to be in the normal form

$$\zeta = P\mathbf{z},$$

which is

$$\begin{cases} \dot{\zeta}_1 = -\cos\left(\zeta_1 + \zeta_2\right) + 1 + v \\ \dot{\zeta}_2 = -\zeta_2 - \cos\left(\zeta_1 + \zeta_2\right) + 1 + v. \end{cases}$$

To this end, the following theorem applies.

Theorem 5.1 (Normal Form Theorem). *Suppose that, in the normal form*

$$\begin{cases} \dot{\zeta}_1 = f(v, \zeta_1, \zeta_2) \\ \dot{\zeta}_2 = -\zeta_2 + g(v, \zeta_1, \zeta_2), \end{cases} \tag{5.4}$$

the two nonlinear functions f and g are both nontrivial functions of v. There are three cases:

(i) *if*

$$\frac{\partial f}{\partial v}(0,0,0) \neq 0 \qquad \text{and} \qquad \frac{\partial^2 f}{\partial \zeta_1^2}(0,0,0) \neq 0$$

then there is a saddle-node bifurcation at $v_0 = 0$;

(ii) *if*

$$v \frac{\partial f}{\partial v}(0,0,0) \cdot \frac{\partial^2 f}{\partial \zeta_1^2}(0,0,0) < 0$$

then there are two hyperbolic equilibria: one is a saddle node and the other is a stable node;

(iii) *if*

$$v \frac{\partial f}{\partial v}(0,0,0) \cdot \frac{\partial^2 f}{\partial \zeta_1^2}(0,0,0) > 0$$

then there is no equilibrium point (so, no bifurcation).

Proof. See [Wiggins (1990)]: p. 216. $\qquad\square$

Note that, since nonsingular linear transforms do not change the system qualitative behaviors (topological properties), the type of bifurcation concluded by the theorem can be transformed back to the original system without altering its qualitative properties and dynamical behaviors. The only corresponding changes are the state variables and the parameter value.

Example 5.7. Return to Example 5.6 on the controlled pendulum. It can be easily verified that

$$\frac{\partial f}{\partial v}(0,0,0) = 1 \neq 0$$

$$\frac{\partial^2 f}{\partial \zeta_1^2}(0,0,0) = - \cos(\zeta_1 + \zeta_2)\Big|_{\zeta_1=\zeta_2=0} = -1 \neq 0.$$

Therefore, at $v_0 = 0$ the system in the normal form has a saddle-node bifurcation. After transforming back to the original system, it has a saddle-node bifurcation at $u_0 = 1$.

Note that one may also exchange ζ_1 and ζ_2 to obtain a "dual theorem" for a "dual system" of the normal form (5.4). This may be useful in some cases.

5.3.2 *Decoupled 2-Dimensional Systems*

Some 2-dimensional (i.e. planar) systems can be decoupled, so that the bifurcation analysis becomes much easier.

Consider a 2-dimensional parameterized system,

$$\begin{cases} \dot{x} = f(x, y; \mu) \\ \dot{y} = g(x, y; \mu), \end{cases}$$

which has a bifurcation point $(x^*, y^*, \mu_0) = (0, 0, 0)$. Applying the Taylor expansion at $(x^*, y^*) = (0, 0)$, one can rewrite the system as

$$\begin{bmatrix} \dot{x} \\ \dot{y} \end{bmatrix} = \begin{bmatrix} a(\mu) \\ b(\mu) \end{bmatrix} + J(\mu) \begin{bmatrix} x \\ y \end{bmatrix} + \text{HOT},$$

where J is the Jacobian and HOT represents all the higher-order terms. Since $\mu = \mu_0 = 0$ is a bifurcation value, $a(0) = b(0) = 0$. Let $\lambda_1(\mu)$ and $\lambda_2(\mu)$ be the eigenvalues of $J(\mu)$. Then, using a nonsingular linear transform, one can assume without loss of generality that $J(\mu) = \text{diag}\{\lambda_1(\mu), \lambda_2(\mu)\}$. It then follows that

$$\begin{cases} \dot{x} = a(\mu) + \lambda_1(\mu)\, x + \text{HOT} \\ \dot{y} = b(\mu) + \lambda_2(\mu)\, x + \text{HOT}. \end{cases} \tag{5.5}$$

Theorem 5.2. *Consider the decoupled 2-dimensional system (5.5).*

(i) *If $da(\mu)/d\mu\big|_{\mu=0} \neq 0$, then there exists a single branch of equilibria of the system in a neighborhood of the bifurcation point $(x^*, y^*, \mu_0) = (0, 0, 0)$, which is of saddle-node type.*

(ii) (a) *If $da(\mu)/d\mu\big|_{\mu=0} = 0$ and*

$$\frac{\partial^2 f}{\partial \mu^2}(0,0,0) \cdot \frac{\partial^2 f}{\partial x^2}(0,0,0) - \left[\frac{\partial^2 f}{\partial \mu \partial x}(0,0,0)\right]^2 < 0,$$

then there are two branches of equilibria, which intersect and exchange stability at the bifurcation point, and the bifurcation is either transcritical or of pitchfork type.

(b) *If $da(\mu)/d\mu\big|_{\mu=0} = 0$ and*

$$\frac{\partial^2 f}{\partial \mu^2}(0,0,0) \cdot \frac{\partial^2 f}{\partial x^2}(0,0,0) - \left[\frac{\partial^2 f}{\partial \mu \partial x}(0,0,0)\right]^2 > 0,$$

then the bifurcation point has an isolated point, and actually the only solution of the system in a neighborhood of this point is the point itself.

Proof. See [Glendinning (1994)]: pp. 206–219.   □

It is remarked that one may exchange $x \leftrightarrow y$ and $f \leftrightarrow g$ to have a "dual" theorem.

Example 5.8. Consider the following decoupled 2-dimensional system:

$$\begin{cases} \dot{x} = \mu - x^2 \\ \dot{y} = -y. \end{cases}$$

Observe that the second equation is linear and stable: $y(t) = e^{-t} \to 0$ as $t \to \infty$; the first one is 1-dimensional with a saddle-node bifurcation as seen in Example 5.2. The phase portraits of this system for different values of μ are shown in Fig. 5.11(a). This system satisfies the saddle-node bifurcation condition stated in Theorem 5.2, since

$$a(\mu) = \mu \qquad \text{and} \qquad \frac{da(\mu)}{d\mu}\bigg|_{\mu=0} = 1 \neq 0\,.$$

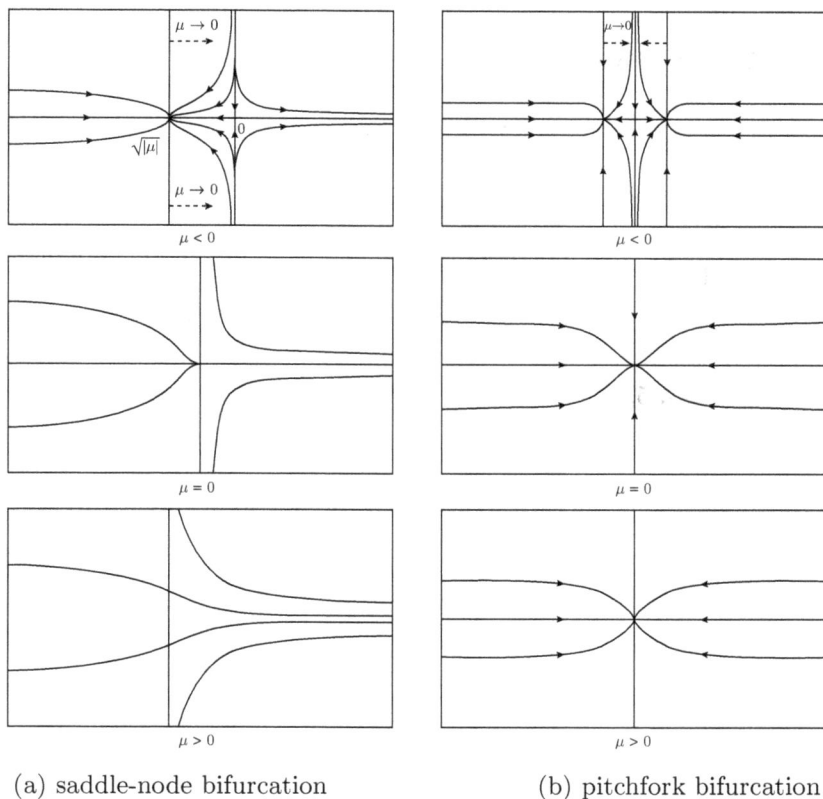

(a) saddle-node bifurcation   (b) pitchfork bifurcation

Fig. 5.11   Pitchfork bifurcation in a 2-dimensional system.

Example 5.9. Consider the following decoupled 2-dimensional system:

$$\begin{cases} \dot{x} = -\mu\,x - x^3 \\ \dot{y} = -y\,. \end{cases}$$

Observe that the second equation is linear and stable: $y(t) = e^{-t} \to 0$ as $t \to \infty$; and the first one is 1-dimensional with a pitchfork bifurcation as seen in Example 5.3. This system satisfies the pitchfork bifurcation condition stated in Theorem 5.2 above, since

$$\frac{\partial^2 f}{\partial \mu^2}(0,0,0) \cdot \frac{\partial^2 f}{\partial x^2}(0,0,0) - \left[\frac{\partial^2 f}{\partial \mu \partial x}(0,0,0)\right]^2 = -1 < 0.$$

The phase portraits of this system for different values of μ are shown in Fig. 5.11(b).

Example 5.10. Consider the Lotka–Volterra system:

$$\begin{cases} \dot{x} = x\,(\mu - x) - (x+1)\,y^2 \\ \dot{y} = y\,(x - 1). \end{cases}$$

Its equilibrium points and curves, when plotted in the $(x, y, \mu := z)$-space, include:

$$(1)\ \begin{cases} x_1^* = 0 \\ y_1^* = 0 \\ z_1^* = \mu \end{cases} \qquad (2)\ \begin{cases} x_2^* = \mu \\ y_2^* = 0 \\ z_2^* = \mu \end{cases} \qquad (3)\ \begin{cases} x_3^* = 1 \\ 2\left(y_3^*\right)^2 = \mu - 1 \\ z_3^* = \mu \end{cases}$$

as shown in Fig. 5.12.

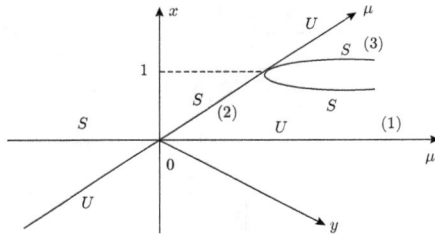

Fig. 5.12   Equilibrium points and curves of the Lotka–Voterra system.

The system Jacobian is

$$J = \begin{bmatrix} \mu - 2x - y^2 & -2(x+1)y \\ y & x - 1 \end{bmatrix}.$$

Case (1). The eigenvalues of J, evaluated at (x_1^*, y_1^*, z_1^*), are $\lambda_1(\mu) = \mu$ and $\lambda_2(\mu) = -1$. So, there is a bifurcation point at $\mu_0 = 0$, which is stable for $\mu < \mu_0 = 0$ (S in Fig. 5.12) but is unstable for $\mu > \mu_0 = 0$ (U in Fig. 5.12).

Case (2). The eigenvalues of J, evaluated at (x_2^*, y_2^*, z_2^*), are $\lambda_1(\mu) = -\mu$ and $\lambda_2(\mu) = \mu - 1$. So, there are two bifurcation points, at $\mu_{01} = 0$ and $\mu_{02} = 1$, respectively; the first one is stable for $\mu_{01} = 0 < \mu < 1$ (S in Fig. 5.12) but the second is unstable for $\mu < \mu_{01} = 0$ and $\mu > \mu_{02} = 1$ (U in Fig. 5.12).

Case (3). The eigenvalues of J, evaluated at (x_3^*, y_3^*, z_3^*), are

$$\lambda_{1,2}(y) = \frac{y^2 - 1 \pm \sqrt{(y^2 - 9)^2 - 80}}{2}$$

or

$$\lambda_{1,2}(\mu) = \frac{\mu - 3 \pm \sqrt{(\mu - 19)^2 - 320}}{4}.$$

There exist bifurcation points only for $\mu \geq 1$, which is stable for $1 \leq \mu < 3$ but is unstable for $\mu > 3$.

It should be noted that, in Case (3) above, the two eigenvalues $\mu_{1,2}(\mu)$ are

(a) both real negative, if $1 < \mu < 19 - \sqrt{320}$;
(b) complex conjugate with negative real parts, if $19 - \sqrt{320} < \mu < 3$;
(c) complex conjugate with positive real parts, if $\mu > 19 + \sqrt{320}$.

Also, as shown in Fig. 5.13,

(a) there is a pitchfork bifurcation at $\mu = 1$;
(b) there is a Hopf bifurcation at $\mu = 3$.

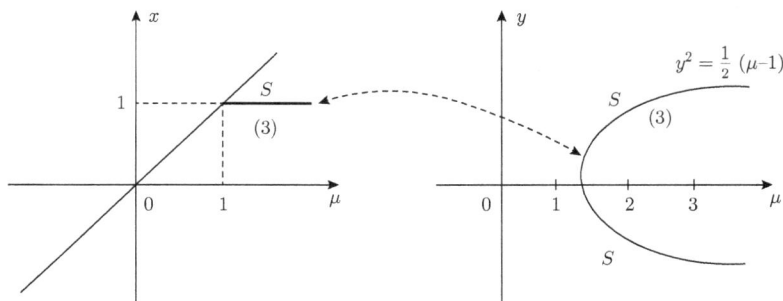

Fig. 5.13   Two other bifurcations of the Lotka–Voterra system.

5.3.3 *Hopf Bifurcation of 2-Dimensional Systems*

For a 2-dimensional non-hyperbolic system, the following is the main result on Hopf bifurcation (see Fig. 5.14).

Theorem 5.3 (Poincaré–Andronov–Hopf Theorem). *Suppose that the 2-dimensional parameterized system*

$$\begin{cases} \dot{x} = f(x, y; \mu) \\ \dot{y} = g(x, y; \mu) \end{cases} \tag{5.6}$$

has a zero equilibrium point, $(x^*, y^*) = (0, 0)$, and that its Jacobian has a pair of complex conjugate eigenvalues, $\lambda(\mu)$ and $\bar{\lambda}(\mu)$. If

$$\frac{d \operatorname{Re}\{\lambda(\mu)\}}{d\mu} \bigg|_{\mu=\mu_0} > 0, \qquad \beta(\mu) \neq 0,$$

for some μ_0, then

   (i) *$\mu = \mu_0$ is a bifurcation point of the system;*
  (ii) *for close enough values $\mu < \mu_0$, the zero equilibrium point is asymptotically stable;*
 (iii) *for close enough values $\mu > \mu_0$, the zero equilibrium point is unstable;*
 (iv) *for close enough values $\mu \neq \mu_0$, the zero equilibrium is surrounded by a limit cycle of magnitude $O(\sqrt{|\mu - \mu_0|})$.*

Proof. See [Arrowsmith and Place (1990)]. □

Example 5.11. Consider the following 2-dimensional system:

$$\begin{cases} \dot{x} = y + x\left(\mu - x^2 - y^2\right) \\ \dot{y} = -x + y\left(\mu - x^2 - y^2\right). \end{cases}$$

Its Jacobian matrix is

$$J = \begin{bmatrix} \mu - 3x^2 - y^2 & 1 - 2xy \\ -1 - 2xy & \mu - x^2 - 3y^2 \end{bmatrix}_{(x^*, y^*)=(0,0)} = \begin{bmatrix} \mu & 1 \\ -1 & \mu \end{bmatrix},$$

with complex conjugate eigenvalues $\lambda_1 = \mu + j$ and $\lambda_2 = \mu - j$, satisfying $\alpha(\mu) = \mu$, $\beta(\mu) = 1 \neq 0$, and $\partial\alpha(\mu)/\partial\mu = 1 > 0$.

Theorem 5.3 implies that, as μ increases to pass the value of $\mu_0 = 0$, the system changes its stability and a limit cycle will emerge with magnitude $O(\sqrt{|\mu - \mu_0|}) = O(\sqrt{|\mu|})$.

Using the polar coordinates, the system can be rewritten as

$$\begin{cases} \dot{\rho} = \rho\left(\mu - \rho^2\right) \\ \dot{\theta} = -1. \end{cases}$$

It is easy to verify, e.g. by computer plots (see Fig. 5.14), that

(i) if $\mu \leq 0$ then the system orbit will spiral inward to the zero equilibrium point (stable focus);

(ii) if $\mu > 0$ then $(0,0)$ becomes unstable, and a limit cycle of radius $\rho_0 = \sqrt{\mu}$ will suddenly emerge as μ is changed from negative to positive through the bifurcation point $\mu = \mu_0 = 0$.

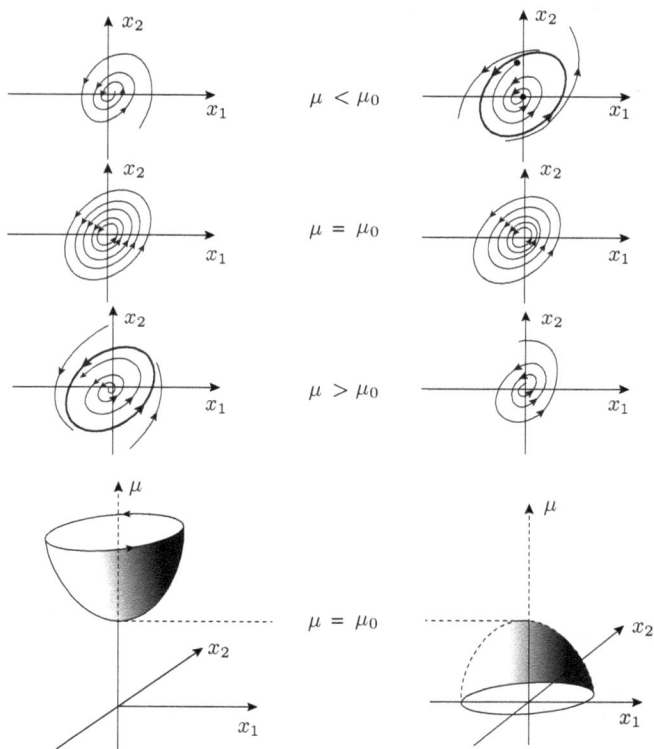

Fig. 5.14   Hopf bifurcation in a 2-dimensional system.

5.4   Poincaré Maps

Consider a general 2-dimensional autonomous system,

$$\begin{cases} \dot{x} = f(x,y) \\ \dot{y} = g(x,y) \end{cases}$$

along with its phase plane. Let Γ be a curve starting from an equilibrium point of the system with the property that it cuts each solution orbit on the phase plane transversely, namely, not tangential to the orbit anywhere.

Consider a point, $P_0 = P_0(x_0, y_0)$, on Γ. Suppose that an orbit passing through Γ at P_0 returns to Γ after some time, being cut by Γ again but probably at a different point, $P_1 = P_1(x_1, y_1)$, as illustrated in Fig. 5.15. This new point, P_1, is called the *first return point* and the map

$$M_\Gamma: \quad (x_0, y_0) \to (x_1, y_1) \qquad (\text{or} \quad M_\Gamma: \quad P_0 \to P_1)$$

is called the *Poincaré map*, denoted as

$$(x_1, y_1) = M_\Gamma(x_0, y_0) \qquad (\text{or} \quad P_1 = M_\Gamma(P_0)).$$

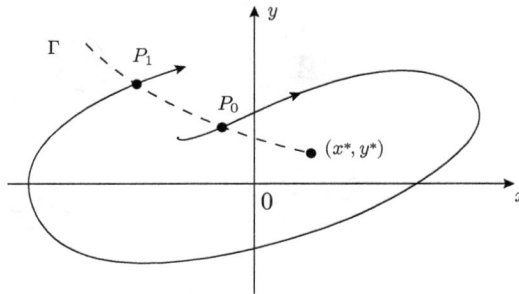

Fig. 5.15   The Poincaré first return map.

If the orbit continues to make a second turn and cuts Γ at $P_2 = P_2(x_2, y_2)$, then

$$(x_2, y_2) = M_\Gamma(x_1, y_1) = M_\Gamma\big(M_\Gamma(x_0, y_0)\big) = M_\Gamma^2(x_0, y_0).$$

After n returns, one has

$$(x_n, y_n) = M_\Gamma^n(x_0, y_0) \qquad (\text{or} \quad P_n = M_\Gamma^n(P_0)),$$

as visualized in Fig. 5.16.

In the situation shown as in Fig. 5.16, since $P^* = M_\Gamma^n(P^*)$ for all n, P^* is an equilibrium point on a periodic orbit. Moreover, as illustrated by the figure, since

$$P_0 \to P_1 \to \cdots \to P^* \qquad \text{and} \qquad P_0' \to P_1' \to \cdots \to P^*,$$

this periodic orbit is a stable limit cycle. This implies that if a Poincaré map M_Γ for a given autonomous system can be appropriately selected, and a point satisfying $P^* = M_\Gamma(P^*)$ exists such that

$$M_\Gamma^n(P_0) \to P^* \qquad \text{and} \qquad M_\Gamma^n(P_0') \to P^*,$$

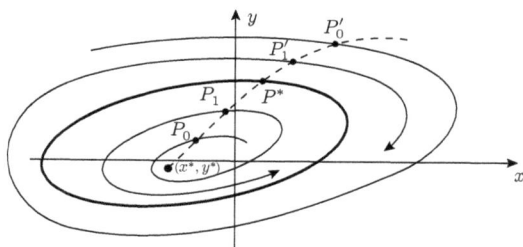

Fig. 5.16 The Poincaré map with multiple returns.

where P_0 and P_0' are located on two opposite sides of P^* at its located solution curve, then it can be concluded that the system has a stable limit cycle. The following example shows more details.

Example 5.12. Consider the following 2-dimensional parameterized system:

$$\begin{cases} \dot{x} = \mu\,x + y - x\sqrt{x^2 + y^2} \\ \dot{y} = -x + \mu\,y - y\sqrt{x^2 + y^2}\,, \end{cases}$$

which has an equilibrium point $(x^*, y^*) = (0,0)$. Consider an orbit $\Gamma : x > 0, y = 0$, namely, the positive semi-x-axis.

In polar coordinates, this system is written as

$$\begin{cases} \dot{r} = r\,(\mu - r) \\ \dot{\theta} = -1\,, \end{cases}$$

which has a solution

$$\begin{cases} r = \mu\,r_0 \,/\, (r_0 + (\mu - r_0)\,e^{-\mu t}) \\ \theta = -t + \theta_0\,, \end{cases}$$

or, simply,

$$r = \frac{\mu\,r_0}{r_0 + (\mu - r_0)\,e^{-\mu(\theta - \theta_0)}}\,.$$

Note that, in polar coordinates, the first return is completed for a 2π-change of θ and, for this example, it is -2π. With $\theta_0 = 0$ and for $0 \le \mu \le r_0$, the Poincaré map is given by

$$r_1 = M_\Gamma(r_0) = \frac{\mu\,r_0}{r_0 + (\mu - r_0)\,e^{-\mu\,2\pi}}\,.$$

Therefore, for $P_0 = (r_0, 0)$, one has

$$r_n = M_\Gamma^n(r_0) = \frac{\mu\,r_0}{r_0 + (\mu - r_0)\,e^{-\mu\,2n\pi}}\,,$$

and $\theta_n = 0$ (which, actually, can be arbitrary). Thus, if $r_0 = \mu$ and $\theta_0 = 0$ are chosen, then $r_n \to \mu$ and $\theta_n \to 0$ as $n \to \infty$. This implies that $P^* = (\mu, 0)$, as shown in Fig. 5.17, where the stable limit cycle is $r = \mu$, which passes the point P^*.

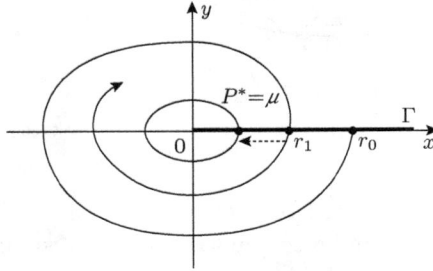

Fig. 5.17   Poincaré map and the stable limit cycle in Example 5.12.

5.5   Strange Attractors and Chaos

Return to the general higher-dimensional autonomous system
$$\dot{\mathbf{x}} = \mathbf{f}(\mathbf{x}), \qquad \mathbf{x}(t_0) = \mathbf{x}_0 \in R^n, \tag{5.7}$$
and recall the concept of ω-limit sets (or positive limit sets) introduced in Definition 2.4: a set of points $\mathbf{z} \in R^n$ is an ω-*limit set* of system (5.7), if there is a solution orbit $\mathbf{x}(t)$ of the system such that
$$||\mathbf{x}(t_n) - \mathbf{z}|| \to 0 \qquad \text{as} \quad n \to \infty \text{ and } t_n \to \infty.$$

It is clear that all stable equilibrium points and all periodic solution orbits (e.g. limit cycles), which need not be stable, are positive limit sets; for example, every solution orbit of the linear harmonic oscillator described by $\dot{x} = y$ and $\dot{y} = -x$ is a positive limit set.

Note that the difference between a stable equilibrium point and a limit cycle as a positive limit set is that the former does not include the trajectory that approaches the limit but the latter does. In the former case, the system solution orbit is called a *manifold*, which is formally defined as follows.

Definition 5.2. Let $\mathbf{x}^*$ be a saddle-node type of equilibrium point. The stable manifold of $\mathbf{x}^*$, denoted $W^s(\mathbf{x}^*)$, is the set of points $\mathbf{x}$ satisfying that the system solution orbit $\varphi_t(\mathbf{x}_0)$ approaches $\mathbf{x}^*$ as $t \to \infty$. An unstable manifold of $\mathbf{x}^*$, denoted $W^u(\mathbf{x}^*)$, is the set of points $\mathbf{x}$ satisfying that the system solution orbit $\varphi_t(\mathbf{x}_0)$ approaches $\mathbf{x}^*$ as $t \to -\infty$.

The concepts of stable and unstable manifolds are illustrated in Fig. 5.18.

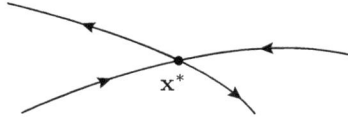

Fig. 5.18 Stable and unstable manifolds of a saddle-node equilibrium point.

Example 5.13. Consider a 3-dimensional linear system with real eigenvalues, $\lambda_1 > 0 > \lambda_2 > \lambda_3$. A solution of the system is given by

$$\varphi_t(\mathbf{x}_0) = c_1 e^{\lambda_1 t} \mathbf{w}_1 + c_2 e^{\lambda_2 t} \mathbf{w}_2 + c_3 e^{\lambda_3 t} \mathbf{w}_3 \,,$$

where $\mathbf{w}_i$ are the corresponding eigenvectors and c_i are constants determined by the initial conditions, $i = 1, 2, 3$.

This system has a unique ω-limit set, which is the zero equilibrium point (the origin). The stable and unstable manifolds are determined by its eigenvalues in this linear case: the stable manifold is the hyperplane spanned by $\mathbf{w}_2$ and $\mathbf{w}_3$, and the unstable manifold is the line curve determined by $\mathbf{w}_1$. These two manifolds intersect transversally at the zero equilibrium point, as illustrated in Fig. 5.19.

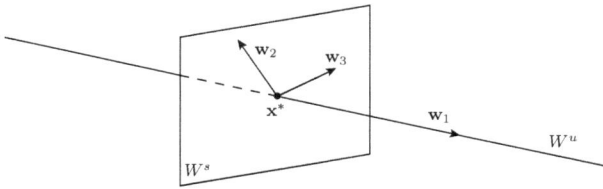

Fig. 5.19 Stable and unstable manifolds of an equilibrium point in a 3-dimensional linear system.

Now, return to the notion of positive limit sets that are not necessarily single points.

Definition 5.3. Consider the general autonomous system (5.7). A set Λ inside a region $\Omega \subseteq R^n$ is called an *attractor* of the system, if the positive limit set of an arbitrary solution orbit of the system lies in Ω then it must lie in Λ.

Intuitively, by definition, an attractor is a limit set and, within some region containing this limit set, all system solution trajectories tend to evolve toward it.

As a couple of simple examples, a stable focus point is an attractor, for which the region $\Omega \subseteq R^n$ is the basin of attraction (see Definition 2.4), and the limit cycle of the van der Pol oscillator (see Fig. 1.12) is an attractor, for which the region is $\Omega = R^2 \backslash \{0\}$.

Note that in a general situation, Ω in Definition 5.3 can be a set but need not be a region. The largest possible set Ω is the basin of attraction containing the set Λ. For example, in the special case where $\Lambda = \{0\}$, this reduces to the basin of attraction of the zero equilibrium point.

Note also that an attractor can be a point, a finite set of points, a curve, a manifold, or even a complicated set with a fractal structure, which is referred to as a *strange attractor*.

5.5.1   *Chaotic Lorenz System*

Typically, chaotic systems have strange attractors. Before introducing the concept and notion of *chaos*, a representative example is discussed.

Example 5.14. Consider the Lorenz system

$$
\begin{cases}
\dot{x} = \sigma \left(y - x \right) \\
\dot{y} = - xz + r\,x - y \\
\dot{z} = xy - b\,z \,,
\end{cases}
\tag{5.8}
$$

where σ, r, and b are positive constants. It is easy to verify that this system has three equilibrium points:

$$
(1)\ \begin{cases} x_1^* = 0 \\ y_1^* = 0 \\ z_1^* = 0\,, \end{cases}
\quad (2)\ \begin{cases} x_2^* = \sqrt{b(r-1)} \\ y_2^* = \sqrt{b(r-1)} \\ z_2^* = r-1\,, \end{cases}
\quad (3)\ \begin{cases} x_3^* = -\sqrt{b(r-1)} \\ y_3^* = -\sqrt{b(r-1)} \\ z_3^* = r-1\,. \end{cases}
$$

Here, by examining the eigenvalues of the system Jacobian at the three equilibrium points, it can be verified that the first equilibrium point is stable if $r < 1$ but unstable if $r > 1$, while the stability of the other two equilibrium points depends on the values of all three parameters σ, r, and b; for example, they are both unstable if $r = \sigma(\sigma + b + 3)/(\sigma - b - 1) = 24.74 \cdots$.

In the following discussions, select and fix $\sigma = 10$ and $b = \frac{8}{3}$, and gradually change the values of r as a parameter:

Case 1. $0 < r < 1$.

In this case, (x_1^*, y_1^*, z_1^*) is a global stable focus point. The other equilibrium points, (x_2^*, y_2^*, z_2^*) and (x_3^*, y_3^*, z_3^*), are complex conjugates, so they cannot be displayed in the phase space.

Case 2. $1 < r < 24.74 \cdots$.

In this case, (x_1^*, y_1^*, z_1^*) becomes unstable, with other two real equilibrium points (x_2^*, y_2^*, z_2^*) and (x_3^*, y_3^*, z_3^*) bifurcating out, which are both local stable point attractors. The phase portraits when $1 < r < 13.926 \cdots$ and $13.926 \cdots < r < 24.74 \cdots$ are shown in Figs. 5.20(a) and 5.20(b), respectively.

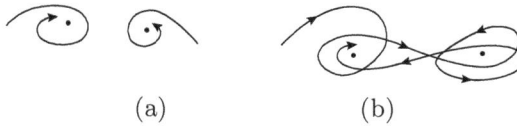

(a)   (b)

Fig. 5.20 The two local point attractor of the Lorenz system.

Case 3. $r = 24.74 \cdots$.

In this case, both (x_2^*, y_2^*, z_2^*) and (x_3^*, y_3^*, z_3^*) become unstable and a Hopf bifurcation appears.

Case 4. $r = 28$.

This is the most interesting situation where chaos emerges: the system orbit is spiraling around one of the two equilibrium points, (x_2^*, y_2^*, z_2^*) and (x_3^*, y_3^*, z_3^*), for a certain period of time (which is unpredictable beforehand), then suddenly jumps to the vicinity of another equilibrium point, which it spirals around for a while (again, the time period of such encircling is unpredictable beforehand), and then it suddenly switches back to the first equilibrium point, $\cdots$. This switching process continues indefinitely and infinitely, but the system orbit never converges to nor diverges from either equilibrium point, and never exactly repeats itself. Thus, the system orbit spiraling around the two equilibrium points virtually approach a strange attractor. This wandering system orbit is said to be a *chaotic orbit*, while the Lorenz system is called a *chaotic system*. This *chaotic attractor* of the Lorenz system is visualized in Fig. 5.21.

The concept of chaos is rather difficult to precisely define in rigorous mathematical terms, due to its highly complicated nature and multiple complex characteristics. Nevertheless, a simple, easily-understandable,

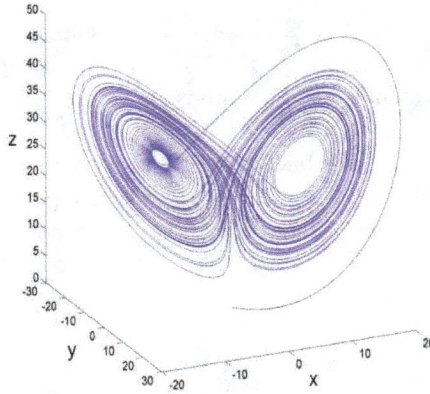

Fig. 5.21   The strange attractor of the Lorenz system.

convenient way to describe and conceptually correct, although not precisely rigorous "definition" of chaos can be given, as follows.

Definition 5.4. A phase orbit is said to be *chaotic*, if it is bounded in a region confined within the phase space but is not convergent (to a point) and is not periodic (not even quasi-periodic).

Here, a system orbit is said to be *quasi-periodic* if it is not-exactly but irregularly similar to a periodic orbit. It is different from the so-called *almost periodic* orbit that has increasing regularity over time.

5.5.2   *Poincaré–Bendixson Theorem*

As mentioned above, a chaotic orbit must be bounded in the phase space. This is intuitively necessary; otherwise, the orbit diverges so it will disappear from the phase space. However, boundedness is obviously not sufficient; for instance, an orbit spiraling toward a stable focus is bounded but it is certainly not chaotic. The following is an important result on the consequence of boundedness for 2-dimensional (planar) systems.

Theorem 5.4 (Poincaré–Bendixson Theorem).   *Consider a 2-dimensional autonomous system, $\dot{\mathbf{x}} = \mathbf{f}(\mathbf{x})$, $\mathbf{x}(t_0) = \mathbf{x}_0 \in R^n$, defined on a bounded closed region $\Omega \subset R^n$. Let Γ be a solution orbit of the system, which enters region Ω and then stays inside forever. In this case, the orbit Γ*

must be one of the following types: (i) *it is a closed orbit;* (ii) *it approaches a closed orbit;* (iii) *it approaches an equilibrium point.*

Proof. See [Verhulst (1996)]: p. 47.   □

These three possibilities are illustrated in Fig. 5.22.

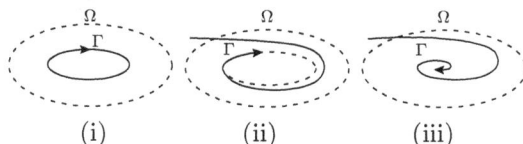

Fig. 5.22   Three cases of a bounded orbit on the plane.

Corollary 5.1. *For a continuous-time autonomous system to have chaos, its dimension has to be at least three.*

5.5.3   *Some Other Chaotic Systems*

Note that the Poincaré–Bendixson Theorem does not apply to nonautonomous systems since a 2-dimensional nonautonomous system may have chaos. Two typical examples are the van der Pol oscillator (1.24) and the Duffing oscillator (1.30), as discussed in more detail below.

Example 5.15. The van der Pol oscillator is described by

$$\begin{cases} \dot{x} = x - \frac{1}{3}x^3 - y + p + q\,\cos(\omega t) \\ \dot{y} = c\,(x + a - b\,y)\,, \end{cases} \tag{5.9}$$

which is chaotic when

$$(p,\, q,\, a,\, b,\, c\omega) = (0.0,\, 0.74,\, 0.7,\, 0.8,\, 0.1,\, 1.0)\,,$$

with the strange attractor shown in Fig. 5.23.

Example 5.16. The Duffing oscillator is described by

$$\begin{cases} \dot{x} = y \\ \dot{y} = -a\,x - b\,x^3 - c\,y + q\,\cos(\omega t)\,, \end{cases} \tag{5.10}$$

which is chaotic when

$$(a,\, b,\, c,\, q,\, \omega) = (-1.1,\, 1.0,\, 0.4,\, 1.8,\, 1.8)\,,$$

with the strange attractor shown in Fig. 5.24.

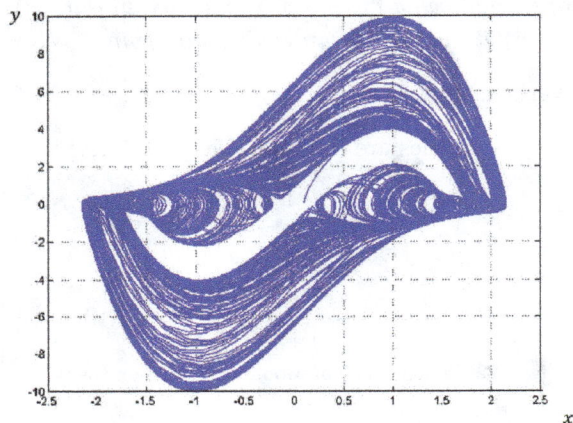

Fig. 5.23   The van der Pol chaotic attractor.

Note also that the Poincaré–Bendixson Theorem does not apply to autonomous systems of dimension three or higher since they may have chaos. Simple examples of 3-dimensional autonomous systems that have only quadratic nonlinearity but can generate chaos include the Lorenz system (see Example 5.14), Chua's circuit (see Fig. 1.26), and the Rössler and Chen systems to be discussed below (see, e.g. [Chen (2020)]).

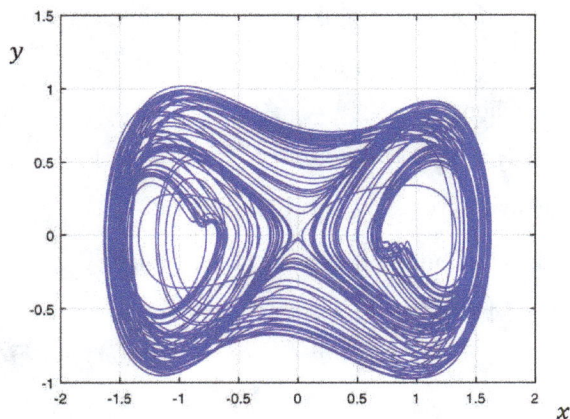

Fig. 5.24   The Duffing chaotic attractor.

Example 5.17. The Rössler system

$$\begin{cases} \dot{x} = -y - z \\ \dot{y} = x + a\,y \\ \dot{z} = x\,z - b\,z + c, \end{cases} \tag{5.11}$$

is chaotic when $(a,\,b,\,c) = (0.2,\,5.7,\,0.2)$, with the strange attractor shown in Fig. 5.25.

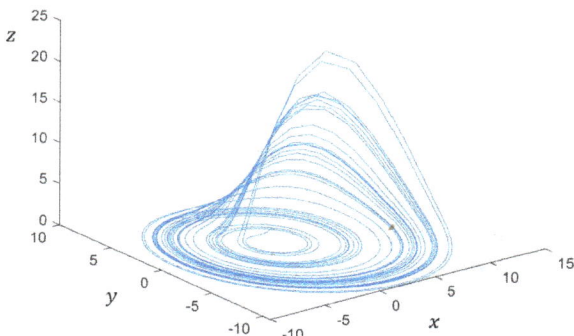

Fig. 5.25   The Rössler chaotic attractor.

Example 5.18. The Chen system

$$\begin{cases} \dot{x} = a\,(y - x) \\ \dot{y} = (c - a)\,x - x\,z + c\,y \\ \dot{z} = x\,y - b\,z \end{cases} \tag{5.12}$$

is chaotic when $(a,\,b,\,c) = (35,\,3,\,28)$, with the strange attractor shown in Fig. 5.26.

It is important to clarify that although the system equations of (5.12) are quite similar to that of the Lorenz system (5.8), they are topologically non-equivalent; namely, there is no diffeomorphism (e.g. nonsingular coordinates transform) that can transfer one to another [Chen (2020)].

Note that although a strange attractor is an indication of chaos, which may not be easy to find in general, however, there are some other characteristics of chaos that can be relatively easily verified, even before a strange attractor is found, which are introduced below.

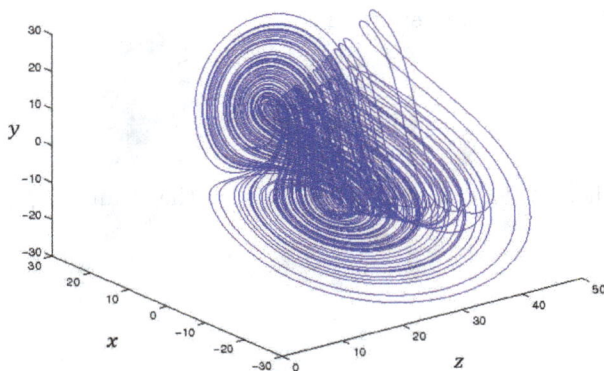

Fig. 5.26   The Chen chaotic attractor.

Note also that for higher-dimensional continuous-time autonomous systems, the Poincaré–Bendixson Theorem may actually be further extended so as to include at least one more possibility from the three possibilities (i)–(iii) stated in Theorem 5.4, for example, the following:

(iv) *The system orbit* Γ *approaches a limit set.*

Here, a limit set includes a strange attractor because a strange attractor is an ω-limit set (see Definition 2.3). Note, however, that there are some other limit sets that are not equilibrium points, limit cycles, or strange attractors. Also, depending on the definition of chaos, a strange attractor can be non-chaotic. In this regard, usually, a *chaotic attractor* is referred to as a strange attractor for a system that is *sensitive to initial conditions*, which is discussed next.

5.5.4   *Characterizations of Chaos*

Regarding the possibility (iv) above, when Γ approaches a chaotic attractor, locally it must have the following features: on the one hand, it keeps approaching a certain subset (e.g. a point), meaning that it possesses some sort of "negative Jacobian eigenvalue" along that particular direction of approaching, on the average; on the other hand, it keeps leaving this subset (or point), meaning that it possesses a "positive Jacobian eigenvalue" along that direction (see Fig. 5.18). Usually, when a system has a positive Jacobian eigenvalue, it will diverge along the direction specified by the corresponding eigenvector. However, the global boundedness property of the system prohibits its global divergence. It is easy to imagine that this global boundedness is determined by the summation of all Jacobian eigenvalues, which should be negative. Thus, a suitable combination of these attracting and repelling features of the strange attractor leads the system orbits to very complicated dynamical behaviors such as chaos.

5.5.4.1 *Lyaponov Exponent: Continuous-Time Systems*

For a nonlinear autonomous system, to quantify the average of a Jacobian eigenvalue of the system, which is evaluated at different operating points throughout the entire dynamical process, the concept of *Lyapunov exponent* is useful.

For a 1-dimensional autonomous system,

$$\dot{x} = f(x), \qquad x(t_0) = x_0,$$

defined on a domain $\mathcal{D} \in R$, if for almost all initial conditions $x_0 \in \mathcal{D}$ (i.e. except perhaps a set of measure zero), the solution $x(t) = x(t; x_0)$ of the system behaves like

$$|x(t; x_0)| \to e^{\lambda t} \qquad \text{for large enough } t > t_0,$$

where $\lambda = \lambda(x_0)$ can be evaluated by

$$\lambda(x_0) = \lim_{t \to \infty} \frac{1}{t} \ln |x(t; x_0)|.$$

This $\lambda(x_0)$, if it is finite, is called the *Lyapunov exponent* of the system orbit $x(t; x_0)$ starting from x_0.

Obviously, the concept of Lyapunov exponent is a generalization of eigenvalue for linear systems. It is also clear that the Lyapunov exponent is sensitive to the system's initial conditions $x_0 \in \mathcal{D}$, which will be further discussed below in Definition 5.6.

For a higher-dimensional autonomous system,

$$\dot{\mathbf{x}} = \mathbf{f}(\mathbf{x}), \qquad \mathbf{x}(t_0) = \mathbf{x}_0 \in R^n,$$

the maximum Lyapunov exponent is defined by

$$\lambda(x_0) = \lim_{t \to \infty} \frac{1}{t} \ln ||\mathbf{z}(t; \mathbf{x}_0)||, \tag{5.13}$$

where $\mathbf{z}(t) = \mathbf{z}(t; \mathbf{x}_0)$ is a solution of the corresponding linearized system

$$\begin{cases} \dot{\mathbf{z}} = J(\mathbf{x})\mathbf{z} \\ \mathbf{z}(t_0) = \mathbf{x}_0, \end{cases}$$

in which $J(\mathbf{x}) = \partial \mathbf{f}(\mathbf{x})/\partial \mathbf{x}$ is the system Jacobnian with $\mathbf{x} = \mathbf{x}(t; \mathbf{x}_0)$.

Table 5.2 Lyapunov exponents of some typical chaotic systems.

System	Lyapunov Exponent
Chen	$(\lambda_1, \lambda_2, \lambda_3) = (1.983,\ 0.000,\ -11.986)$
Chua	$(\lambda_1, \lambda_2, \lambda_3) = (0.230,\ 0.000,\ -1.7800)$
Lorenz	$(\lambda_1, \lambda_2, \lambda_3) = (0.897,\ 0.000,\ -14.565)$
Rössler	$(\lambda_1, \lambda_2, \lambda_3) = (0.130,\ 0.000,\ -14.100)$

For a 3-dimensional autonomous system to have chaos, it is common that its three Lyapunov exponents are $\lambda_1 > 0$, $\lambda_2 = 0$, and $\lambda_3 < 0$, denoted as $(+, 0, -)$, with $\lambda_1 + \lambda_3 < 0$. The Lyapunov exponents for several typical chaotic systems are listed in Table 5.2.

In general, for 3-dimensional systems, the three Lyapunov exponents $(-, -, -)$ correspond to an equilibrium point, $(0, -, -)$ a periodic orbit, $(0, 0, -)$ a torus, and $(+, 0, -)$ chaos as shown in Table 5.2.

It is remarked that to characterize chaos, in addition to the Lyapunov exponent there are some other measures, such as positive entropy, fractal dimension, and continuous power spectrum [Wiggins (1990); Hoppensteadt (2000)], which are not further studied in this text.

5.5.4.2 *Lyaponov Exponent: Discrete-Time Systems*

Lyapunov exponents in discrete-time systems are quite different, although the concept and therefore the definition are similar.

For a discrete-time "autonomous" system,

$$\mathbf{x}_{k+1} = \mathbf{f}\left(\mathbf{x}_k\right), \qquad \mathbf{x}_0 \in R^n\,,$$

denote its Jacobian at the kth step by

$$J_k = J_k(\mathbf{x}_0) = \frac{\partial \mathbf{f}\left(\mathbf{x}_k\right)}{\partial \mathbf{x}_k}\,,$$

which all depend on the initial state $\mathbf{x}_0$ since $\mathbf{x}_k = \mathbf{f}^k(\mathbf{x}_0)$. Let

$$P_k = P_k(\mathbf{x}_0) := J_k J_{k-1} \cdots J_2 J_1$$

be the kth product of such Jacobians, and let $e_i = e_i(\mathbf{x}_0)$ be the eigenvalues of P_k, arranged according to

$$|e_1| \geq |e_2| \geq \cdots \geq |e_n| \geq 0\,.$$

The ith Lyapunov exponent of the system orbit $\{\mathbf{x}_k\}$, starting from $\mathbf{x}_0$, is defined by

$$\lambda_i(\mathbf{x}_0) = \lim_{k \to \infty} \frac{1}{k} \ln |e_i(\mathbf{x}_0)|, \qquad i = 1, \ldots, n. \tag{5.14}$$

Differing from the continuous-time setting, a discrete-time system can be chaotic even if it is 1-dimensional and has (the only) one positive Lyapunov exponent. The logistic map (5.3) is a typical example of this type:

$$x_{k+1} = \mu\, x_k (1 - x_k), \qquad x_0 \in (0, 1), \quad 0 < \mu \le 4, \tag{5.15}$$

This map is chaotic when, say, $\mu = 4.0$ (or slightly smaller). In this case, its only Lyapunov exponent is $\lambda = \ln 2 = 0.693 \cdots$. Its diagram of period-doubling bifurcation leading to chaos is shown in Fig. 5.8.

Another example is the 2-dimensional Hénon map,

$$\begin{cases} x_{k+1} = 1 - a\, x_k^2 + y_k \\ y_{k+1} = b\, x_k, \end{cases} \tag{5.16}$$

which is chaotic when $(a, b) = (1.4, 0.3)$, with Lyapunov exponents $\lambda_1 = 0.603$ and $\lambda_2 = -2.34$ without a zero exponent. A typical chaotic phase portrait of the Hénon map is shown in Fig. 5.27.

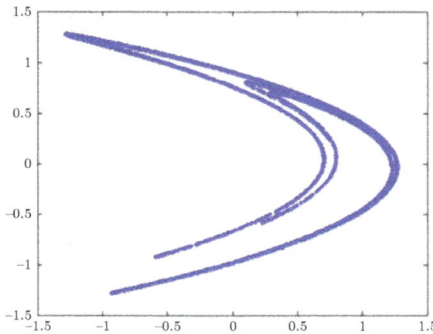

Fig. 5.27   The Hénon chaotic attractor.

In both continuous-time and discrete-time autonomous systems, a positive maximum Lyapunov exponent is a necessary condition for chaos to exist.

5.6   Chaos in Discrete-Time Systems

Chaos in discrete-time systems, or maps, has a rather precise definition. For a 1-dimensional map, the following definition [Touhey (1997)] is convenient to use and verify.

Definition 5.5. A map $M : S \to S$, where S is a nonempty set in a bounded and compact domain, is *chaotic* if and only if for any two nonempty open subsets, U and V, of S, there exist a point $x_0 \in S$ and a period-n orbit of f, with a non-negative integer n, such that $f(x_0) \cap U \neq \emptyset$ and $f(x_0) \cap V \neq \emptyset$.

This definition is equivalent to another, more common definition of discrete chaos given earlier *in the sense of Devaney* [Devaney (1987)].

Definition 5.6. A map $M : S \to S$, where S is a nonempty set in a bounded and compact domain, is *chaotic* if

  (i) it is sensitive to initial conditions;
 (ii) it is topologically transitive;
(iii) it has a dense set of periodic orbits.

Here, the meaning of each condition is further interpreted as follows:

(i) "Sensitivity to initial conditions" means that starting from two arbitrarily close initial points, x_1 and x_2, in S, the corresponding orbits, $M^n(x_1)$ and $M^n(x_2)$, will fall far apart after a large enough number of iterations, n. For instance, let $S = [a, b]$ be a bounded interval in R_+, and $M : S \to S$ be a map. This property of sensitivity to initial conditions implies that for any prescribed $\delta \in [0, b - a]$, as long as $x_1 \neq x_2$ in S, no matter how close they are, after a large enough number of iterations, n, one has $\lVert M^n(x_1) - M^n(x_2) \rVert > \delta$. This property is illustrated in Fig. 5.28.

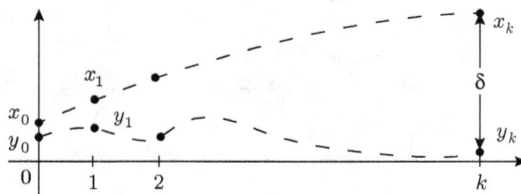

Fig. 5.28   Sensitivity to initial conditions.

(ii) "Topological transitivity" means that for any nonempty open set $\Omega \subseteq S$, no matter how small it is, an orbit of the map will sooner or later travel into it. This property is illustrated in Fig. 5.29.

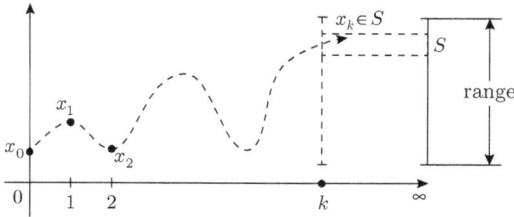

Fig. 5.29   Topological transitivity property.

(iii) "Dense set of periodic orbits" means that the map has infinitely many periodic orbits of different periods, and all these periodic orbits constitute a dense set in S. The period-doubling bifurcation diagram of the logistic map, shown in Fig. 5.8, with $\mu \in (0, 4.0]$, best illustrates this property.

Example 5.19. Consider again the logistic map

$$x_{k+1} = 4\,x_k\left(1 - x_k\right).$$

First, it can be verified that this map is equivalent to

$$x_{k+1} = 2\,x_k \qquad (\text{mod } 1)\,, \tag{5.17}$$

since they both have the same (unique) solution

$$x_k = \frac{1}{2}\left(1 - \cos(2\pi 2^k y_0)\right),$$

where

$$y_0 = \frac{1}{2\pi}\cos^{-1}\left(1 - 2\,x_0\right).$$

Then, one can show that the map (5.17) satisfies conditions (i)–(iii) in Definition 5.6, as verified below:

(i) System (5.17) has a Lyapunov exponent $\lambda = \ln 2 > 0$. Indeed,

$$\lambda = \lim_{k\to\infty} \frac{1}{k}\ln\left|J_k J_{k-1}\cdots J_2 J_1\right| = \lim_{k\to\infty} \frac{1}{k}\ln 2^k = \ln 2\,,$$

where $J_k = 2$ for all $k = 1, 2, \ldots$, and the modulo operation does not change the derivative, except at the turning points, which does not affect the concerned derivatives. Therefore, the map is sensitive to initial conditions because its orbit is diverging.

(ii) Because of the mod-1 operation, the map (5.17) is equivalent to the following double angle map from the unit circle to itself:

$$\angle x_{k+1} = 2\angle x_k \quad (\text{mod } 2\pi), \tag{5.18}$$

where $\angle$ is the angle, and x_0 is any initial point on the unit circle, as shown in Fig. 5.30.

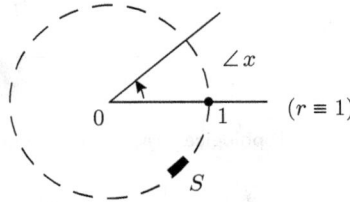

Fig. 5.30   The double angle map from the unit circle to itself.

Since the angle is doubled on each iteration, and the number 2 is not an integer multiplier of π, it is clear that for any nonempty open arc-segment S on the circle, sooner or later (i.e. there is an index k), the point will fall into the arc-segment: $x_k \in S$. Therefore, this double angle map is topologically transitive, so is the map (5.17).

(iii) Since Eq. (5.18) is 2π-periodic, one has

$$\angle x_{k+1+n} = 2^n \angle x_k \pm 2m\pi, \quad m, n, k = 0, 1, 2, \dots.$$

Thus, by letting $k \to \infty$, one can see that for each given pair (n, m), this equation yields an equilibrium point, $\angle x^*$, satisfying

$$\angle x^* = 2^n \angle x^* \pm 2m\pi, \quad m, n, k = 0, 1, 2, \dots.$$

This gives

$$\angle x^* = \frac{\pm 2m\pi}{2^n - 1}, \quad m = 0, 1, 2, \dots, 2^n - 1; \; n = 1, 2, \dots.$$

Therefore, the map has infinitely many periodic orbits of different periods, all located on the unit circle. Moreover, as $n, m \to \infty$, one can see that all these periodic points become dense on the unit circle, almost uniformly, so does the map (5.17).

For a higher-dimensional map: $M : S \to S$, where S is a bounded and compact set in R^n with $n > 1$, there is a definition of chaos introduced in [Marotto (1978)], which generalizes an earlier definition *in the sense of*

Li–Yorke [Li and Yorke (1975)] from 1-dimensional to higher-dimensional maps.

Consider an n-dimensional discrete-time autonomous system,

$$\mathbf{x}_{k+1} = \mathbf{f}(\mathbf{x}_k), \tag{5.19}$$

where $\mathbf{f}$ is a vector-valued continuous nonlinear function. Let $\mathbf{f}'(\mathbf{x})$ and $\det(\mathbf{f}'(\mathbf{x}))$ be the Jacobian of $\mathbf{f}$ at $\mathbf{x}$ and its determinant, respectively, and let $B_r(\mathbf{x})$ be a closed ball of radius $r > 0$ centered at $\mathbf{x}$ in R^n.

Definition 5.7. An equilibrium point $\mathbf{x}^*$ of (5.19), assuming it exists, is said to be a *snapback repeller* if

(i) there exists a real number, $r > 0$, such that $\mathbf{f}$ is differentiable with all eigenvalues of $\mathbf{f}'(\mathbf{x})$ exceeding unity in absolute value for all $\mathbf{x} \in B_r(\mathbf{x}^*)$;
(ii) there exists an $\mathbf{x}^0 \in B_r(\mathbf{x}^*)$, with $\mathbf{x}^0 \neq \mathbf{x}^*$, such that for some positive integer m, $\mathbf{f}^m(\mathbf{x}^0) = \mathbf{x}^*$ and $\mathbf{f}^m(\mathbf{x}^0)$ are differentiable at $\mathbf{x}^0$ with $\det((\mathbf{f}^m)'(\mathbf{x}^0)) \neq 0$.

The concept of snapback repeller is illustrated in Fig. 5.31.

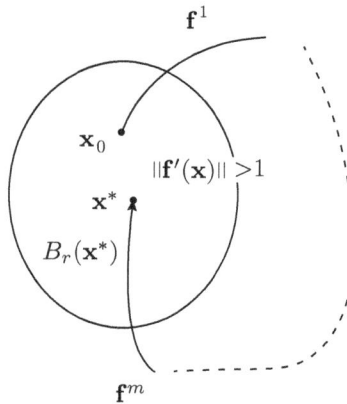

Fig. 5.31   The snapback repeller illustrated.

Theorem 5.5 (Marotto Theorem).   *If system* (5.19) *has a snapback repeller then the system is chaotic in the sense of Li and Yorke, namely,*

(i) *there exists a positive integer n such that, for every integer $p \geq n$, system* (5.19) *has p-periodic points;*

(ii) *there exist a scrambled set (i.e. an uncountable invariant set) S containing no periodic points, such that*

(a) $\mathbf{f}(S) \subset S$,

(b) *for every $\mathbf{y} \in S$ and any periodic point $\mathbf{x}$ of (5.19):*

$$\limsup_{k \to \infty} \left\| \mathbf{f}^k(\mathbf{x}) - \mathbf{f}^k(\mathbf{y}) \right\| > 0,$$

(c) *for every $\mathbf{x}, \mathbf{y} \in S$ with $\mathbf{x} \neq \mathbf{y}$,*

$$\limsup_{k \to \infty} \left\| \mathbf{f}^k(\mathbf{x}) - \mathbf{f}^k(\mathbf{y}) \right\| > 0;$$

(iii) *there exists an uncountable subset S_0 of S such that, for any $\mathbf{x}, \mathbf{y} \in S_0$,*

$$\liminf_{k \to \infty} \left\| \mathbf{f}^k(\mathbf{x}) - \mathbf{f}^k(\mathbf{y}) \right\| = 0.$$

Exercises

5.1 For each of the following equations, determine the type of bifurcation when μ is varied, and sketch the corresponding bifurcation diagram:

$$\dot{x} = 1 + \mu x + x^2, \qquad \dot{x} = \mu x + x^2,$$
$$\dot{x} = x + \mu x^3, \qquad \dot{x} = \mu - 3 x^2.$$

5.2 For each of the following equations, determine the type of bifurcation when μ is varied, and sketch the corresponding bifurcation diagram:

$$\dot{x} = \mu x - x(1 - x), \qquad \dot{x} = x - \mu x(1 - x),$$
$$\dot{x} = x - \mu x^3, \qquad \dot{x} = \mu x - \frac{x}{1 + x}.$$

5.3 Sketch all the periodic solutions of the following system, and indicate their stabilities:

$$\ddot{x} + \dot{x}\,(x^2 + \dot{x}^2 - 1) + x = 0.$$

5.4 Determine the limit cycles and their stabilities for the following two systems:

$$\begin{cases} \dot{\rho} = \rho\,(1 - \rho^2)\,(9 - \rho^2) \\ \dot{\theta} = 1 \end{cases}$$

and

$$\begin{cases} \dot{\rho} = \rho\,(\rho - 1)\,(\rho^2 - 2) \\ \dot{\theta} = 1. \end{cases}$$

5.5 For the following systems, discuss their bifurcations and sketch their phase portraits as μ varies:

$$\begin{cases} \dot{x} = \mu x - x^2 \\ \dot{y} = -y \end{cases} \quad \text{and} \quad \begin{cases} \dot{x} = \mu x + x^3 \\ \dot{y} = -y. \end{cases}$$

5.6 For the following system, discuss its bifurcation and sketch its phase portrait as μ varies:

$$\begin{cases} \dot{x} = y - 2x \\ \dot{y} = \mu + x^2 - y. \end{cases}$$

5.7 Consider the system

$$\begin{cases} \dot{x} = \mu x - y + x y^2 \\ \dot{y} = x + \mu y + y^3. \end{cases}$$

Study its Hopf bifurcation as μ is varied to pass $\mu_0 = 0$. Is this bifurcation supercritical or subcritical?

5.8 Consider the following biased van der Pol oscillator:

$$\ddot{x} + \mu\left(x^2 - 1\right)\dot{x} + x = c,$$

where c is a constant. Find the curve in the μ–c plane on which Hopf bifurcations occur.

5.9 Consider the following system:

$$\begin{cases} \dot{x} = \left(\frac{1}{\sqrt{2}}z + \frac{\alpha}{2}\right)y - \beta x \\ \dot{y} = \left(\frac{1}{\sqrt{2}}z - \frac{\alpha}{2}\right)x - \beta y \qquad (\alpha, \beta > 0). \\ \dot{z} = \sqrt{2}\left(1 - xy\right) \end{cases}$$

Use a typical quadratic Lyapunov function to argue that in a large enough neighborhood of $(0,0,0)$ there must be an attractor, and this attractor is not $(0,0,0)$.

5.10 Consider the skew tent map

$$x_{k+1} = \begin{cases} x/a & \text{if} \quad 0 \le x \le a \\ (1-x)/(1-a) & \text{if} \quad a < x \le 1, \end{cases}$$

with initial state $x_0 \in (0,1)$ and parameter $0.5 \le a < 1$.

(i) Pick arbitrarily x_0 and a to generate a relatively long time series of the map (e.g. 100,000 points). Then, calculate its Lyapunov exponent λ. Verify if your result satisfies

$$0 < -\ln\left(a\right) \le \lambda \le -\ln\left(1 - a\right).$$

(ii) Pick an arbitrary x_0 with a fixed $a = 0.5$. Find all the equilibrium points of the map and classify their stabilities.

(iii) Play around, with different values of x_0 and $a \in [0.5, 1)$, to see if you can get period-2, period-3, period-4 orbits, and even chaos.

Chapter 6

Nonlinear Systems Control

This chapter briefly discusses the basic ideas, principles and general methods for designing controllers for nonlinear systems, including chaotic systems.

It should be made clear at the start that this chapter is not about specific engineering design of any kind of controller with implementation guidelines, but rather, only some mathematical descriptions of approaches and methodologies, mostly based on the theories and techniques developed in the previous chapters.

6.1 Feedback Control of Nonlinear Systems

Consider a general nonlinear dynamical system,

$$\begin{cases} \dot{\mathbf{x}}(t) = \mathbf{f}(\mathbf{x}, \mathbf{u}, t) \,, \\ \mathbf{y}(t) = \mathbf{h}(\mathbf{x}, \mathbf{u}, t) \,, \end{cases} \tag{6.1}$$

where $\mathbf{x}(t)$ is the system state vector, $\mathbf{y}(t)$ the output vector, and $\mathbf{u}(t)$ the control input vector. Here, in a general discussion, once again it is assumed that all the necessary conditions on the vector-valued functions $\mathbf{f}$ and $\mathbf{h}$ are satisfied such that the system has a unique solution in a certain bounded region of the state space for each given initial state $\mathbf{x}_0 = \mathbf{x}(t_0)$, with initial time $t_0 \geq 0$.

Given a reference signal, $\mathbf{r}(t)$, which can be either a constant vector (setpoint) or a function of time (target trajectory), the problem is to design a controller in the state-feedback form

$$\mathbf{u}(t) = \mathbf{g}(\mathbf{x}, t) \,, \tag{6.2}$$

or, sometimes, in the output-feedback form

$$\mathbf{u}(t) = \mathbf{g}(\mathbf{y}, t) \,,$$

where $\mathbf{g}$ is a nonlinear (including linear) vector-valued function, such that the controlled system

$$\begin{cases} \dot{\mathbf{x}}(t) = \mathbf{f}\big(\mathbf{x}, \mathbf{g}(\mathbf{x}, t), t\big), \\ \mathbf{y}(t) = \mathbf{h}\big(\mathbf{x}, \mathbf{g}(\mathbf{x}, t)\big), \end{cases} \tag{6.3}$$

can be driven by the feedback control $\mathbf{g}(\mathbf{x}, t)$ to track the target:

$$\lim_{t \to t_f} \|\mathbf{y}(t) - \mathbf{r}(t)\| = 0, \tag{6.4}$$

where the terminal time $t_f \leq \infty$ is predetermined according to the application, typically $t_f = \infty$ in theory, which means that the time is long enough in practice, and $\|\cdot\|$ is the Euclidean norm of a vector.

The special case with $\mathbf{r}(t) \equiv \mathbf{c}$, a constant vector, is sometimes referred to as a stabilization problem and, without loss of generality, one may set $\mathbf{c} = 0$ by a change of variables in the nonlinear systems setting.

Since the second equation in system (6.1) is merely a mapping, which can be easily handled in general, it is ignored in the following discussion by simply setting $\mathbf{h}(\cdot) = I$, the identity mapping, so that $\mathbf{y} = \mathbf{x}$, without further discussion on output-feedback control in this chapter.

6.1.1 *Engineering Perspectives on Controllers Design*

From the general mathematical form of the nonlinear controller (6.2), it might seem that this controller can be arbitrary in any form and of any complexity. However, it is very important to point out that in a state-feedback controller design, particularly in finding a nonlinear controller for a given nonlinear system, one must bear in mind that the controller should be (much) simpler than the given system to be practical, implementable and cost-effective.

One might easily have a false impression that a nonlinear controller could be in any formulation since it is nonlinear anyway. For example, to design a nonlinear controller $\mathbf{u}(t)$ to drive the state vector $\mathbf{x}(t)$ of the nonlinear system

$$\dot{\mathbf{x}}(t) = \mathbf{f}(\mathbf{x}(t), t) + \mathbf{u}(t)$$

to track a target trajectory $\mathbf{x}^*(t)$, one might attempt to use

$$\mathbf{u}(t) = \dot{\mathbf{x}}^*(t) - \mathbf{f}(\mathbf{x}(t), t) + K\big(\mathbf{x}(t) - \mathbf{x}^*(t)\big),$$

where K has all its eigenvalues with negative real parts. This controller leads to the tracking-error dynamical system

$$\dot{\mathbf{e}}(t) = K\,\mathbf{e}(t), \qquad \text{with} \quad \mathbf{e}(t) = \mathbf{x}(t) - \mathbf{x}^*(t),$$

yielding $\mathbf{e}(t) \to 0$, or $\mathbf{x}(t) \to \mathbf{x}^*(t)$, as $t \to \infty$.

Is there anything wrong mathematically? No.

Another example is: for a given nonlinear controlled system in the canonical form

$$
\begin{cases}
\dot{x}_1(t) = x_2(t) \\
\dot{x}_2(t) = x_3(t) \\
\quad \vdots \\
\dot{x}_{n-1}(t) = x_n(t) \\
\dot{x}_n(t) = f\big(x_1(t), \cdots, x_n(t)\big) + u(t),
\end{cases}
$$

suppose that one wants to find a nonlinear controller $u(t)$ to drive the state vector $\mathbf{x}(t) = \big[x_1(t) \ \cdots \ x_n(t)\big]^{\top}$ to a target state, $\widetilde{\mathbf{x}}$, i.e.

$$
\mathbf{x}(t) \to \widetilde{\mathbf{x}} \qquad \text{as} \quad t \to \infty.
$$

It is mathematically straightforward to use the controller

$$
u(t) = -f\big(x_1(t), \cdots, x_n(t)\big) + k_{\mathrm{c}}\big(x_n(t) - \widetilde{x}_n\big)
$$

with an arbitrary constant $k_{\mathrm{c}} < 0$. This controller yields

$$
\dot{x}_n(t) = k_{\mathrm{c}}\big(x_n(t) - \widetilde{x}_n\big),
$$

which guarantees that $e_n(t) := x_n(t) - \widetilde{x}_n \to 0$ as $t \to \infty$ since $\dot{e}_n(t) = k_{\mathrm{c}} e_n(t)$ and $k_{\mathrm{c}} < 0$. Indeed, the resulting n-dimensional controlled system is a completely controllable linear system, $\dot{\mathbf{x}} = A\mathbf{x} + \mathbf{b}u$, with

$$
A = \begin{bmatrix} 0 & 1 & 0 & \cdots & 0 \\ 0 & 0 & 1 & \cdots & 0 \\ \vdots & \vdots & \vdots & \ddots & 1 \\ 0 & 0 & 0 & \cdots & 0 \end{bmatrix} \qquad \text{and} \qquad \mathbf{b} = \begin{bmatrix} 0 \\ \vdots \\ 0 \\ 1 \end{bmatrix}.
$$

Therefore, a suitable constant control gain exists for the state-feedback controller, $u = k_{\mathrm{c}} x_n$, to achieve that $\mathbf{x}(t) \to \widetilde{\mathbf{x}}$ as $t \to \infty$.

Is there anything wrong mathematically? No.

Just to show one more example, suppose that one would like to find a nonlinear controller, say $\mathbf{u}_k$ in the discrete-time setting, to drive the state vector $\mathbf{x}_k$ of a given nonlinear control system of the form

$$
\mathbf{x}_{k+1} = \mathbf{f}_k(\mathbf{x}_k) + \mathbf{u}_k
$$

to a target trajectory satisfying a predetermined target dynamical form of $\mathbf{x}_{k+1} = \phi_k(\mathbf{x}_k)$, then mathematically it is very easy to use

$$
\mathbf{u}_k = \phi_k(\mathbf{x}_k) - \mathbf{f}_k(\mathbf{x}_k),
$$

which will bring the original system state $\mathbf{x}_k$ to the target trajectory in one step.

Is there anything wrong mathematically? No.

To this end, all such (exaggerated) examples seem to suggest a "universal controller design methodology" that works for any given system with any given target, and the argument is mathematically correct. However, as all practitioners know, a fatal problem with such a "design" is that the controller removes part of the given system, or is even more complicated than the given system, which has no practical value: every term in a mathematical model of a practical system has a specific physical meaning, representing a resistor, a spring, a neuron, etc., which is a part of the given circuit, machine or bio-system. They should be controlled but not be removed by the designed controller. For instance, if a term R represents a resistor in the mathematical model of a circuit device, then cancelling this R by the controller simply means to remove the resistor from the given device.

Recall that the conventional linear PID (proportional-integral-derivative) controllers never remove any part of the given system (usually called a plant), and this is why they can be commonly used in the industry.

Furthermore, in practice, due to the ubiquitous noise, uncertainty, and parameter mismatch, a precise term-cancellation is generally impossible. As a result, the consequent analysis and implementation of the above cancellation-based design cannot be carried out.

Therefore, a nonlinear controller design is expected to present a simplest possible and cost-effective controller: if a linear controller can be found to do the job, use a linear controller; otherwise, try a simple nonlinear controller with a piecewise or a quadratic nonlinearity and so on. Also, usually full state feedback information is not available in practice, so one should try to design a controller using only output feedback (i.e. partial state feedback), which means the second equation of (6.1) is essential, although this notion will not be further discussed in this chapter. Whether or not one can find a simple, easily implementable, low-cost, and effective controller for a given nonlinear system requires both good theoretical knowledge and design experience.

6.1.2   *A General Approach to Controllers Design*

Returning to the central theme of feedback control for the general nonlinear dynamical system (6.1)–(6.4), a basic idea is first outlined here for tracking control.

Let the target trajectory (or set-point) be $\widetilde{\mathbf{x}}(t)$, and assume that it is differentiable, so that by denoting

$$\dot{\widetilde{\mathbf{x}}}(t) = \mathbf{z}(t), \qquad (6.5)$$

one can subtract this equation (6.5) from the first equation of (6.3), to obtain the error dynamical system

$$\dot{\mathbf{e}} = \mathbf{F}(\mathbf{e}, t), \qquad (6.6)$$

where $\mathbf{e} = \mathbf{x} - \widetilde{\mathbf{x}}$ and

$$\mathbf{F}(\mathbf{e}, t) := \mathbf{f}\big(\mathbf{x}, \mathbf{g}(\mathbf{x}, t), t\big) - \mathbf{z}(t).$$

If the target trajectory $\widetilde{\mathbf{x}}$ is a periodic orbit of the given system (6.1), that is, if it satisfies

$$\dot{\widetilde{\mathbf{x}}} = \mathbf{f}\big(\widetilde{\mathbf{x}}, 0, t\big), \qquad (6.7)$$

then similarly a subtraction of (6.7) from the first equation of (6.1) gives

$$\dot{\mathbf{e}} = \mathbf{F}(\mathbf{e}, t), \qquad (6.8)$$

where $\mathbf{e} = \mathbf{x} - \widetilde{\mathbf{x}}$ and

$$\mathbf{F}(\mathbf{e}, t) := \mathbf{f}\big(\mathbf{x}, \mathbf{g}(\mathbf{x}, t), t\big) - \mathbf{f}\big(\widetilde{\mathbf{x}}, 0, t\big).$$

In either case, e.g. in the second case, which is more difficult in general, the goal of design is to determine the controller $\mathbf{u}(t) = \mathbf{g}(\mathbf{x}, t)$ such that

$$\lim_{t \to \infty} \|\mathbf{e}(t)\| = 0, \qquad (6.9)$$

which implies that tracking control is achieved:

$$\lim_{t \to \infty} \|\mathbf{x}(t) - \widetilde{\mathbf{x}}(t)\| = 0. \qquad (6.10)$$

It is now clear from (6.9) and (6.10) that if zero is a fixed point of the nonlinear system (6.8), then the original controllability problem has been converted to an asymptotic stability problem of this fixed point. Thus, the Lyapunov first and second methods may be applied or modified to obtain rigorous mathematical analysis for the controller's design. This is further discussed in more detail below.

6.2 Feedback Controllers for Nonlinear Systems

This section discusses how a linear or nonlinear controller may be designed for controlling a nonlinear dynamical system based on the rigorous Lyapunov function methods.

6.2.1 *Linear Controllers for Nonlinear Systems*

Given the Lyapunov first method for nonlinear autonomous systems and, if applicable, the linear stability theory for nonlinear nonautonomous systems with weak nonlinearities (see Sec. 3.1), we show that a linear feedback controller may be able to control a nonlinear dynamical system in a rigorous way.

Take the nonlinear, actually chaotic Chua circuit shown in Fig. 1.26 as an example. This circuit is a simple yet very interesting electrical system that displays many typical bifurcation and chaotic phenomena.

Example 6.1. The Chua circuit is shown in Fig. 6.1 again. It consists of one inductor L, two capacitors C_1 and C_2, one linear resistor R, and one nonlinear resistor g, which is a nonlinear function of the voltage across its two terminals: $g = g\big(V_{C_1}(t)\big)$.

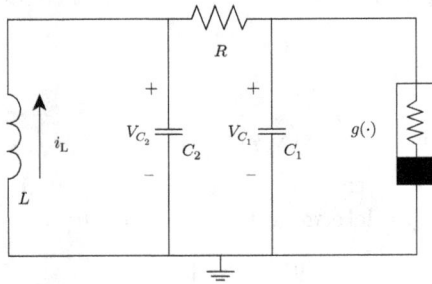

Fig. 6.1   The Chua circuit.

Let $i_L(t)$ be the current through the inductor L, and $V_{C_1}(t)$ and $V_{C_2}(t)$ be the voltages across C_1 and C_2, respectively, as shown in the circuit diagram. Then, it follows from Kirchhoff's laws that

$$C_1 \frac{d}{dt} V_{C_1}(t) = \frac{1}{R} \left[V_{C_2}(t) - V_{C_1}(t) \right] + g\big(V_{C_1}(t)\big),$$

$$C_2 \frac{d}{dt} V_{C_2}(t) = \frac{1}{R} \left[V_{C_1}(t) - V_{C_2}(t) \right] + i_L(t),$$

$$L \frac{d}{dt} i_L(t) = - V_{C_2}(t).$$

In the discussion of controlling the circuit behavior, it turns out to be easier to first apply the nonlinear transformation

$$x_1(\tilde{t}) = V_{C_1}(t), \quad x_2(\tilde{t}) = V_{C_2}(t), \quad x_3(t) = R\,i_L(t), \quad \tilde{t} = t/(C_2 R)$$

to reformulate the circuit equations to the following dimensionless form:

$$\begin{cases} \dot{x} = p\,[\,-x + y - f(x)\,] \\ \dot{y} = x - y + z \\ \dot{z} = -q\,y\,, \end{cases} \tag{6.11}$$

where $p > 0$, $q > 0$, and $f(x) = R\,g(x)$ is a nonlinear function of the form

$$f(x) = m_0\,x + \frac{1}{2}\,(m_1 - m_0)\,(\,|x + 1| - |x - 1|\,)\,,$$

in which $m_0 < 0$ and $m_1 < 0$.

It is known that, with $p = 10.0$, $q = 14.87$, $m_0 = -0.68$, and $m_1 = -1.27$, the circuit displays a chaotic attractor (see Fig. 6.2) along with a limit cycle of large magnitude. This limit cycle is a large unstable periodic orbit, hence unable to be shown in the simulation figure, which encompasses the non-periodic attractor. It is generated due to the eventual passivity of the transistors.

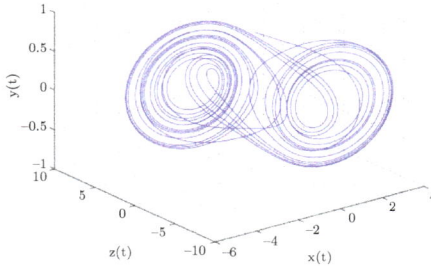

Fig. 6.2   The double scroll chaotic attractor of the Chua circuit.

Now, let $(\bar{x}, \bar{y}, \bar{z})$ be the unstable periodic orbit of the Chua circuit (6.11). Then, the circuit trajectory (x, y, z) can be driven from any (finite) state to reach this periodic orbit by a simple linear feedback control of the form

$$\begin{bmatrix} u_1 \\ u_2 \\ u_3 \end{bmatrix} = -K \begin{bmatrix} x - \bar{x} \\ y - \bar{y} \\ z - \bar{z} \end{bmatrix} = - \begin{bmatrix} k_{11} & 0 & 0 \\ 0 & k_{22} & 0 \\ 0 & 0 & k_{33} \end{bmatrix} \begin{bmatrix} x - \bar{x} \\ y - \bar{y} \\ z - \bar{z} \end{bmatrix} \tag{6.12}$$

with

$$k_{11} \geq -pm_1\,, \qquad k_{22} \geq 0\,, \qquad \text{and} \qquad k_{33} \geq 0\,,$$

where the control can be applied to the trajectory at any time.

A mathematical verification is given as follows. First, one observes that the controlled circuit is

$$\begin{cases} \dot{x} = p\left(-x + y - f(x)\right) - k_{11}\left(x - \bar{x}\right), \\ \dot{y} = x - y + z - k_{22}\left(y - \bar{y}\right), \\ \dot{z} = -q\,y - k_{33}\left(z - \bar{z}\right). \end{cases} \qquad (6.13)$$

Since the unstable periodic orbit $(\bar{x}, \bar{y}, \bar{z})$ is itself a (periodic) solution of the circuit, one has

$$\begin{cases} \dot{\bar{x}} = p\left(-\bar{x} + \bar{y} - f(\bar{x})\right), \\ \dot{\bar{y}} = \bar{x} - \bar{y} + \bar{z}, \\ \dot{\bar{z}} = -q\,\bar{y}, \end{cases} \qquad (6.14)$$

so that a subtraction of (6.14) from (6.13), with the new notation

$$X = x - \bar{x}, \qquad Y = y - \bar{y}, \qquad \text{and} \qquad Z = z - \bar{z},$$

yields

$$\begin{cases} \dot{X} = p\left(-X + Y - \widetilde{f}(x, \bar{x})\right) - k_{11}X, \\ \dot{Y} = X - Y + Z - k_{22}Y, \\ \dot{Z} = -q\,Y - k_{33}Z, \end{cases} \qquad (6.15)$$

where

$$\widetilde{f}(x, \bar{x}) = \begin{cases} m_0(x - \bar{x}) & x \geq 1, \bar{x} \geq 1 \\ m_0 x - m_1\bar{x} + m_1 - m_0 & x \geq 1, -1 \leq \bar{x} \leq 1 \\ m_0(x - \bar{x}) + 2(m_1 - m_0) & x \geq 1, \bar{x} \leq -1 \\ m_1 x - m_0\bar{x} - m_1 + m_0 & -1 \leq x \leq 1, \bar{x} \geq 1 \\ m_1(x - \bar{x}) & -1 \leq x \leq 1, -1 \leq \bar{x} \leq 1 \\ m_1 x - m_0\bar{x} + m_1 - m_0 & -1 \leq x \leq 1, \bar{x} \leq -1 \\ m_0(x - \bar{x}) - 2(m_1 - m_0) & x \leq -1, \bar{x} \geq 1 \\ m_0 x - m_1\bar{x} - m_1 + m_0 & x \leq -1, -1 \leq \bar{x} \leq 1 \\ m_0(x - \bar{x}) & x \leq -1, \bar{x} \leq -1 \end{cases}$$

with $m_1 < m_0 < 0$.

Consider the following Lyapunov function for system (6.15):

$$V(X, Y, Z) = \frac{q}{2}X^2 + \frac{pq}{2}Y^2 + \frac{p}{2}Z^2.$$

It is clear that $V(0, 0, 0) = 0$ and $V(X, Y, Z) > 0$ for all X, Y, Z that are not simultaneously zero. On the other hand, since $p, q > 0$ and $k_{22}, k_{33} \geq 0$, it follows that

$$\begin{aligned}
\dot{V} &= q\,X\dot{X} + pq\,Y\dot{Y} + p\,Z\dot{Z} \\
&= q\,X\big(-p\,X + p\,Y - p\,\widetilde{f}(x,\bar{x}) - k_{11}X\big) \\
&\quad + pq\,Y\big(X - Y + Z - k_{22}\,Y\big) + p\,Z\big(-q\,Y - k_{33}\,Z\big) \\
&= -pq\,X^2 + 2pq\,XY - pq\,Y^2 - pq\,X\widetilde{f}(x,\bar{x}) \\
&\quad - q\,k_{11}\,X^2 - k_{22}\,pq\,Y^2 - k_{33}\,p\,Z^2 \\
&= -p\big[q\,(X-Y)^2 + q\,k_{22}\,Y^2 + k_{33}\,Z^2\big] - q\big(p\,X\widetilde{f}(x,\bar{x}) + k_{11}\,X^2\big) \\
&\le 0
\end{aligned}$$

for all X, Y, and Z, if

$$p\,X\widetilde{f}(x,\bar{x}) + k_{11}\,X^2 \ge 0 \tag{6.16}$$

for all x and $\bar{x}$. To find the conditions under which (6.16) holds, by a careful examination of the nine possible cases for the function $\widetilde{f}(x,\tilde{x})$ shown above, the following common condition can be obtained:

$$k_{11} \ge \max\{-p\,m_0, -p\,m_1\} = -p\,m_1, \tag{6.17}$$

in which $m_1 < m_0 < 0$. This condition guarantees the inequality (6.16). Hence, if the stated conditions are satisfied, the equilibrium point $(0,0,0)$ of the controlled circuit (6.15) is globally asymptotically stable, so that

$$|X| \to 0, \quad |Y| \to 0, \quad |Z| \to 0 \qquad \text{as} \quad t \to \infty,$$

simultaneously. That is, starting with the feedback control at any time on the chaotic trajectory, one has

$$\lim_{t\to\infty} |x(t) - \bar{x}(t)| = 0, \quad \lim_{t\to\infty} |y(t) - \bar{y}(t)| = 0, \quad \lim_{t\to\infty} |z(t) - \bar{z}(t)| = 0.$$

The tracking result is visualized in Fig. 6.3.

6.2.2 *Nonlinear Controllers for Nonlinear Systems*

It is not always possible to use a linear controller to control a nonlinear dynamical system. In fact, nonlinear feedback controllers are necessary in most cases.

To show one example of a nonlinear controller design, consider the chaotic Duffing oscillator (5.10) again.

Example 6.2. The Duffing oscillator is described by

$$\begin{cases} \dot{x} = y, \\ \dot{y} = -p_2\,x - x^3 - p_1\,y + q\,\cos(\omega t), \end{cases} \tag{6.18}$$

where p_1, p_2, q, and ω are constant parameters.

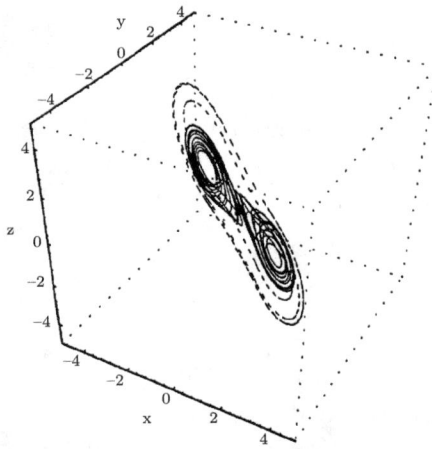

Fig. 6.3   Tracking the unstable limit cycle of the Chua circuit.

With the parameters set $p_1 = 0.4$, $p_2 = -1.1$, $q = 2.1$ (or $q = 1.8$), and $\omega = 1.8$, the Duffing system has a chaotic attractor (see Fig. 5.24). The Duffing system has some inherent unstable limit cycles, which however do not have analytic expressions and even cannot be accurately displayed graphically due to their instability.

For this system, suppose that for controlling its chaotic trajectory to one of its inherent unstable periodic orbits, we design a feedback controller.

Denote by $(\bar{x}, \bar{y}) = (\bar{x}(t), \bar{y}(t))$ the target trajectory, one of its (unknown) unstable periodic orbits. The goal is to control the system trajectory to track this periodic trajectory, namely,

$$\lim_{t \to \infty} |x(t) - \bar{x}(t)| = 0 \quad \text{and} \quad \lim_{t \to \infty} |y(t) - \bar{y}(t)| = 0. \qquad (6.19)$$

Denote the nonlinear feedback controller by $u(t) = h(t; x, \bar{x})$, which is to be determined. By adding the controller to the second equation of the original system, one obtains the following controlled Duffing system:

$$\begin{cases} \dot{x} = y, \\ \dot{y} = -p_2 \, x - x^3 - p_1 \, y + q \, \cos(\omega t) + h(t; x, \bar{x}). \end{cases} \qquad (6.20)$$

Since the periodic orbit $(\bar{x}, \bar{y})$ is itself a solution of the original system, subtracting (6.18), with (x, y) replaced by $(\bar{x}, \bar{y})$ therein, from system (6.20), one obtains the tracking error system

$$\begin{cases} \dot{X} = Y, \\ \dot{Y} = -p_2 \, X - (x^3 - \bar{x}^3) - p_1 \, Y + h(t; x, \bar{x}), \end{cases} \qquad (6.21)$$

where
$$X = x - \bar{x} \quad \text{and} \quad Y = y - \bar{y}.$$
Next, observe that the controlled Duffing system (6.21) is a nonlinear, nonautonomous system; therefore, the Lyapunov first method may not apply. In this particular case, however, the Lyapunov second method can be applied. Indeed, a nonlinear controller $h(x)$ can be designed as follows:
$$h(x) = k\,X + 3\,\bar{x}^2\,X + 3\,\bar{x}\,X^2\,,$$
where k is a constant to be determined. Note that the Duffing system is third-order and nonautonomous but this controller is autonomous and it does not cancel any term of the original system. Consequently, the controlled system (6.21) becomes
$$\begin{cases} \dot{X} = Y \\ \dot{Y} = -\,(k + p_2)\,X - p_1\,Y - X^3\,. \end{cases} \tag{6.22}$$
Consider the Lyapunov function
$$V(X, Y) = \frac{k + p_2}{2}\,X^2 + \frac{1}{4}\,X^4 + \frac{1}{2}\,Y^2\,,$$
which satisfies
$$\begin{aligned} \dot{V} &= (k + p_2)X\dot{X} + X^3\dot{X} + Y\dot{Y} \\ &= (k + p_2)XY + X^3Y + Y[-(k + p_2)X - p_1Y - X^3] \\ &= -p_1 Y^2\,. \end{aligned}$$
It then follows from the LaSalle Invariance Principle (see Theorem 2.10) that if $Y = 0$, but $X \neq 0$, then (6.22) yields
$$\dot{Y} = -(k + p_2)X - X^3\,.$$
It is thus clear that if $k + p_2 > 0$, which gives a criterion for choosing the parameter k, then (i) for $X > 0$, $\dot{Y} < 0$; (ii) for $X < 0$, $\dot{Y} > 0$. Therefore, the system solution trajectory will not stay on the X-axis unless $(X, Y) = (0, 0)$. This means that the zero fixed point of the controlled Duffing system (6.22) is asymptotically stable, so that $X \to 0$ and $Y \to 0$ as $t \to \infty$, namely the goal
$$|x - \bar{x}| \to 0 \quad \text{and} \quad |\dot{x} - \dot{\bar{x}}| \to 0 \quad (t \to \infty)$$
is achieved. Numerically, tracking is achieved with the result visualized in Fig. 6.4.

Finally, it should be emphasized that this simple example is only used to illustrate some basic idea and approach in designing a simple nonlinear controller for a given complex nonlinear system, which is by no means the best possible or most effective design. In fact, in this example, tracking to which periodic trajectory cannot be specified beforehand. Nevertheless, sometimes it would be sufficient if the target is just a periodic trajectory, of whatever period, for which the goal of control has been achieved.

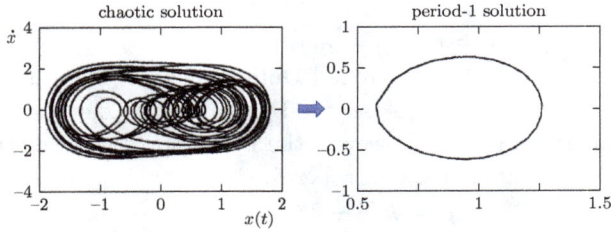

Fig. 6.4 Tracking an unstable limit cycle of the Duffing oscillator.

6.2.3   *General Ideas for Nonlinear Controllers Design*

Consider a general nonlinear and nonautonomous controlled system,

$$\dot{\mathbf{x}} = \mathbf{f}(\mathbf{x}, t) + \mathbf{u}(t), \tag{6.23}$$

which is assumed to possess a periodic orbit $\tilde{\mathbf{x}}$ of period $t_p > 0$: $\tilde{\mathbf{x}}(t + t_p) = \tilde{\mathbf{x}}(t)$, $0 \leq t < \infty$. The goal is to design a feedback controller of the form

$$\mathbf{u}(t) = K\left(\mathbf{x} - \tilde{\mathbf{x}}\right) + \mathbf{g}\left(\mathbf{x} - \tilde{\mathbf{x}}, t\right), \tag{6.24}$$

where K is a constant matrix and $\mathbf{g}$ is a (simple) nonlinear vector-valued function such that the trajectory of the controlled system can track the target periodic orbit $\tilde{\mathbf{x}}$, namely,

$$\lim_{t \to \infty} \|\mathbf{x}(t) - \tilde{\mathbf{x}}(t)\| = 0. \tag{6.25}$$

As usual, the controller (6.24) is added to the given system (6.23) to obtain the controlled system

$$\dot{\mathbf{x}} = \mathbf{f}(\mathbf{x}, t) + \mathbf{u} = \mathbf{f}(\mathbf{x}, t) + K\left(\mathbf{x} - \tilde{\mathbf{x}}\right) + \mathbf{g}\left(\mathbf{x} - \tilde{\mathbf{x}}, t\right). \tag{6.26}$$

Since the target periodic orbit $\tilde{\mathbf{x}}$ is itself a solution of the original system, it satisfies

$$\dot{\tilde{\mathbf{x}}} = \mathbf{f}(\tilde{\mathbf{x}}, \mathbf{t}). \tag{6.27}$$

A subtraction of (6.27) from (6.26) gives

$$\dot{\mathbf{X}} = \mathcal{F}(\mathbf{X}, t) + K\mathbf{X} + \mathbf{g}(\mathbf{X}, t), \tag{6.28}$$

where

$$\mathbf{X} = \mathbf{x} - \tilde{\mathbf{x}} \qquad \text{and} \qquad \mathcal{F}(\mathbf{X}, t) = \mathbf{f}(\mathbf{x}, t) - \mathbf{f}(\tilde{\mathbf{x}}, t).$$

It is clear that $\mathcal{F}(0, t) = 0$ for all $t \in [0, \infty)$.

Now, Taylor-expand the right-hand side of the controlled system (6.28) at $\mathbf{X} = 0$ (i.e. at $\mathbf{x} = \tilde{\mathbf{x}}$), and suppose that the nonlinear controller is designed such that it satisfies $\mathbf{g}(0, t) = 0$. Then, one has

$$\dot{\mathbf{X}} = A(\tilde{\mathbf{x}}, t)\mathbf{X} + \mathbf{h}(\mathbf{X}, K, t), \tag{6.29}$$

where $A(\tilde{\mathbf{x}}, t) = \partial \mathcal{F}(\mathbf{X}, t)/\partial \mathbf{X}\big|_{\mathbf{X}=0}$ and $\mathbf{h}(\mathbf{X}, K, t)$ is the rest of the Taylor expansion, which is a function of t, K, and higher-order terms $o(\mathbf{X})$.

To this end, the design is to determine both the constant control gain matrix K and the nonlinear controller $\mathbf{g}(\mathbf{X}, t)$ based on the linearized model (6.29), such that $\mathbf{X} \to 0$ (i.e. $\mathbf{x} \to \tilde{\mathbf{x}}$) as $t \to \infty$.

The following criteria follow from Theorems 3.1 and 3.4.

Theorem 6.1. *Suppose that, in system* (6.29), $\mathbf{h}(0, K, t) = 0$ *and* $A(\tilde{\mathbf{x}}, t) = A$ *is a constant matrix with all eigenvalues having negative real parts. If*

$$\lim_{||\mathbf{X}|| \to 0} \frac{||\mathbf{h}(\mathbf{X}, K, t)||}{||\mathbf{X}||} = 0$$

uniformly with respect to $t \in [0, \infty)$, *where* $|| \cdot ||$ *is the Euclidean norm, then the controller* $\mathbf{u}(t)$ *defined in* (6.24) *will drive the trajectory* $\mathbf{x}$ *of the controlled system* (6.28) *to the target periodic trajectory* $\tilde{\mathbf{x}}$ *as* $t \to \infty$.

Next, recall the concept of Floquet multipliers for periodic systems introduced in Definition 3.1.

Theorem 6.2. *In system* (6.29), *suppose that* $\mathbf{h}(0, K, t) = 0$ *and that* $\mathbf{h}(\mathbf{X}, K, t)$ *and* $\partial \mathbf{h}(\mathbf{X}, K, t)/\partial \mathbf{X}$ *are both continuous in a bounded region with* $||\mathbf{X}|| < \infty$. *Assume also that*

$$\lim_{||\mathbf{X}|| \to 0} \frac{||\mathbf{h}(\mathbf{X}, K, t)||}{||\mathbf{X}||} = 0,$$

uniformly with respect to $t \in [0, \infty)$. *If all the multipliers of the system* (6.29) *satisfy*

$$|\lambda_i| < 1, \qquad i = 1, \ldots, n, \quad \forall \, t \in [0, \infty),$$

then the nonlinear controller (6.24) *will drive the trajectory* $\mathbf{x}$ *of the original controlled system* (6.29) *to the target periodic trajectory* $\tilde{\mathbf{x}}$ *as* $t \to \infty$.

6.3 More about Nonlinear Controllers Design

6.3.1 *An Illustrative Example*

Consider a class of Liénard equations of the form

$$\ddot{x} + b(\dot{x}) + c(x) = 0, \tag{6.30}$$

which is assumed to have a zero fixed point $(x^*, \dot{x}^*) = (0, 0)$, where $b(\cdot)$ and $c(\cdot)$ are nonlinear functions satisfying

(i) $\dot{x}\, b(\dot{x}) > 0$ for all $\dot{x} \neq 0$;

(ii) $x\, c(x) > 0$ for all $x \neq 0$.

Examples of this type of equations include:

$$\ddot{x} + \dot{x}^3 + x^5 = 0$$

$$\ddot{x} + \frac{\kappa}{m}\, \dot{x}^3 + \frac{g^2}{m\,\ell^2}\, \sin(x) = 0 \qquad (\text{pendumu:} \quad -\pi < x < \pi)$$

$$\ddot{x} - \mu\big(1 - x^2\big)\dot{x} + x = 0 \qquad (\text{van der Pol:} \quad \mu > 0,\ |x| > 1).$$

For this class of systems, an effective Lyapunov function is the total energy function

$$V(x, \dot{x}) = \frac{1}{2}\, \dot{x}^2(t) + \int_0^{x(t)} c(\sigma)\, d\sigma\,.$$

Under conditions (i) and (ii) above, $V(x, \dot{x}) > 0$ for $(x, \dot{x}) \neq (0, 0)$, and

$$\begin{aligned}
\dot{V} &= \dot{x}\,\ddot{x} + c(x)\dot{x} \\
&= -\,\dot{x}\, b(\dot{x}) - \dot{x}\, c(x) + c(x)\,\dot{x} \\
&= -\,\dot{x}\, b(\dot{x}) \\
&< 0 \qquad \text{for all}\ \ \dot{x} \neq 0\,.
\end{aligned}$$

Clearly, the only chance for $\dot{V} = 0$ while $\dot{x} \neq 0$ is when $x \equiv 0$, i.e. on the x-axis. But this axis is not a region, so the Lyapunov instability theorems studied in Sec. 2.4 do not apply. Besides, whenever the system orbit is located on the x-axis, since $\dot{x} \neq 0$, it will leave this axis immediately, and right after that moment, $\dot{V} < 0$, which forces the orbit to move toward the origin. This implies that the zero fixed point of the system is asymptotically stable. If, moreover,

$$\int_0^{x(t)} c(\sigma)\, d\sigma \to \infty \qquad \text{as}\ \ |x| \to \infty\,,$$

then the asymptotic stability is global.

Now, suppose that conditions (i) and (ii) above are not simultaneously satisfied. In this case, the objective is to design a controller, u, to be added to the right-hand side of the given system, namely

$$\ddot{x} + b(\dot{x}) + c(x) = u\,, \tag{6.31}$$

so as to stabilize the zero fixed point.

How can this be accomplished? It is quite natural to try a controller of the form
$$u = u_1(\dot{x}) + u_2(x),$$
where $u_1(\cdot)$ and $u_2(\cdot)$ may be nonlinear, both are to be determined. The controlled system becomes
$$\ddot{x} + \big[\, b(\dot{x}) - u_1(\dot{x}) \,\big] + \big[\, c(x) - u_2(x) \,\big] = 0\,.$$
Thus, it is clear that one has to find u_1 and u_2 such that

(i') $\dot{x}\,[b(\dot{x}) - u_1(\dot{x})] > 0$ for all $\dot{x} \neq 0$;
(ii') $x\,[c(x) - u_2(x)] \geq 0$ for all $x \neq 0$.

The following example illustrates how these conditions can be satisfied in the design.

Example 6.3. Consider the system
$$\ddot{x} - \dot{x}^3 + x^2 = 0\,,$$
which does not simultaneously satisfy conditions (i) and (ii). The objective is to design a controller $u = u_1(\dot{x}) + u_2(x)$ that can ensure both
$$\begin{cases} \dot{x}\,\big[- \dot{x}^3 - u_1(\dot{x})\,\big] > 0 & (\dot{x} \neq 0) \\ x\,\big[x^2 - u_2(x)\,\big] \geq 0 & (x \neq 0)\,. \end{cases}$$
For this purpose, it might be mathematically straightforward to choose
$$u_1(\dot{x}) = -2\,\dot{x}^3 \qquad \text{and} \qquad u_2(x) = x^2 - x\,,$$
which yields
$$u = u_1 + u_2 = -2\,\dot{x}^3 + x^2 - x\,.$$
However, this controller is not desirable since it is even more complicated than the given system; it basically cancels the given nonlinearity and then adds a stable linear portion back to the resulting system. In so doing, it actually replaced the given system by a stable one, which is impractical as discussed in Sec. 6.1.1.

To design a simple and easily implementable controller can sometimes be quite technical. Nevertheless, for the purpose of illustration, a slightly simpler design for this particular example could be
$$u_1(\dot{x}) = -\dot{x}^3 - \dot{x} \qquad \text{and} \qquad u_2(x) = -x\,|x|\,,$$
which ensures that
$$\begin{cases} \dot{x}\,\big[- \dot{x}^3 - u_1(\dot{x})\,\big] = \dot{x}^2 > 0 \\ x\,\big[x^2 - u_2(x)\,\big] = x^2\,\big[x + |x|\,\big] \geq 0\,, \end{cases}$$
for all $x \neq 0$ and $\dot{x} \neq 0$.

6.3.2 *Adaptive Controllers*

The above Lyapunov design method is also useful for controllers design for uncertain systems, where some system parameters are unknown, therefore have to be estimated. The designed controllers using the estimated parameters instead of the unknown parameters, which are not usable, are referred to as *adaptive* controllers.

6.3.2.1 *Observer-Based Adaptive Control*

The following very simple example is used to explain the basic idea and approach.

Example 6.4. Consider an uncertain linear control system,

$$\dot{x} + \alpha\, x = u\,,$$

where α is an unknown constant. The objective is to design a controller, u, such that the controlled state $x(t) \to 0$ as $t \to \infty$.

Case 1.

If $|\alpha| \le \overline{\alpha}$, where the upper bound constant $\overline{\alpha}$ is known, which may be conservative, then the linear controller

$$u = -\,2\,\overline{\alpha}\, x$$

can do the job. Indeed, the controlled system becomes

$$\dot{x} + \left(2\,\overline{\alpha} + \alpha\right) x = 0\,,$$

which has a solution

$$x(t) = x_0\, e^{-(2\overline{\alpha}+\alpha)t} \to 0 \qquad (t \to \infty)\,.$$

Case 2.

If no such an upper bound $\overline{\alpha}$ is known, then usually no linear controller can be designed, since whatever linear controller $u = -kx$ is used, the controlled system orbit will be

$$x(t) = x_0\, e^{-(k+\alpha)t}\,,$$

which does not provide any guideline for the determination of the constant gain k since α is unknown.

In this case, one has to resort to a different methodology. Estimation of α on the line is often necessary, for which the *observer* is a useful tool.

Suppose that $\widehat{\alpha}$ is an estimate of the unknown α, which satisfies a so-called observer equation,

$$\dot{\widehat{x}} = f(x, \widehat{\alpha}),$$

where $f(\cdot, \cdot)$ is to be determined. Use $\widehat{\alpha}$ to design the controller and denote

$$u = u(x, \widehat{\alpha}).$$

To find the observer $f(\cdot, \cdot)$, one may consider the Lyapunov function

$$V(x, \widehat{\alpha}) = \frac{1}{2}x^2 + \frac{1}{2}(\widehat{\alpha} - \alpha)^2,$$

and then look for both $f(\cdot, \cdot)$ and $u(\cdot, \cdot)$ to ensure that

$$\dot{V} \leq -cx^2 < 0$$

for some constant $c > 0$. Here, one may use some class-$\mathcal{K}$ functions (e.g. $\gamma(|x|^2) := c|x|^2$; see Sec. 2.2). Since it is desirable to have

$$\dot{V} = x\dot{x} + (\widehat{\alpha} - \alpha)\dot{\widehat{\alpha}} = x(u - \alpha x) + (\widehat{\alpha} - \alpha)\dot{\widehat{\alpha}} \leq -cx^2,$$

or

$$xu + \widehat{\alpha}\dot{\widehat{\alpha}} - \alpha(x^2 + \dot{\widehat{\alpha}}) \leq -cx^2,$$

one natural choice for the observer equation is

$$\dot{\widehat{\alpha}} = -x^2.$$

Its solution is

$$\widehat{\alpha}(t) = \widehat{\alpha}(0) - \int_0^t x^2(\tau)\,d\tau,$$

where $\widehat{\alpha}(0)$ may be chosen to be zero. Thus, what is wanted becomes

$$\dot{V} = xu - \dot{\widehat{\alpha}}x^2 \leq -cx^2,$$

which suggests the following form for the controller:

$$u = (\widehat{\alpha} - c)x \qquad (c > 0).$$

Usually, it is preferred that the controller can be a negative feedback. For this purpose, c can be any constant satisfying $|\widehat{\alpha}(t)| \leq c$ for all $t \geq 0$.

The entire adaptive control system based on this observer design can be implemented as shown in Fig. 6.5.

The methodology discussed above can be further extended to adaptive control of higher-dimensional uncertain linear systems, and even nonlinear systems. Only the simple single-input single-output (SISO) higher-dimensional linear systems and 1-dimensional nonlinear systems are discussed in the following. The general multiple input multiple out (MIMO) case is not further studied in this textbook.

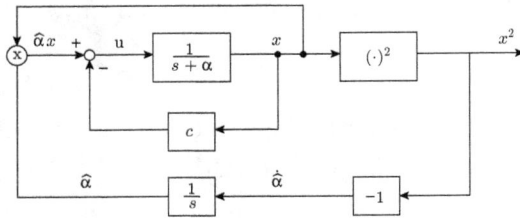

Fig. 6.5   Implementation of the observer-based adaptive control system.

6.3.2.2   *Adaptive Control of Uncertain Linear Systems*

The concept of a positive real transfer function (for SISO systems) and matrix (for MIMO systems) is useful.

Definition 6.1. A transfer function $H(s)$ of an SISO linear system is said to be *positive real* if

$$\text{Re}\{H(s)\} > 0 \qquad \text{for all} \quad \text{Re}\{s\} > 0,$$

and is said to be *strictly positive real* if

$$\text{Re}\{H(s - \varepsilon)\} > 0 \qquad \text{for all} \quad \text{Re}\{s\} \geq \varepsilon > 0.$$

For instance, $H(s) = 1/s$ is positive real but not strictly, and $H(s) = 1/(s + \alpha)$ with $\alpha > 0$ is strictly positive real.

Lemma 6.1 (Kalman–Yakubovich–Popov Lemma).   *Consider the SISO LTI system*

$$\begin{cases} \dot{\mathbf{x}} = A\mathbf{x} + \mathbf{b}u \\ y = \mathbf{c}^\top \mathbf{x}. \end{cases}$$

Its transfer function

$$H(s) = \mathbf{c}^\top [sI - A]^{-1} \mathbf{b}$$

is strictly positive real if and only if there exists positive definite and symmetrical matrices P and Q such that

$$\begin{cases} A^\top P + PA = Q \\ P\mathbf{b} = \mathbf{c}. \end{cases}$$

Proof. See [Khalil (1996)]: pp. 240–241.   □

Theorem 6.3. *Suppose that the SISO LTI system*

$$\begin{cases} \dot{\mathbf{x}} = A\mathbf{x} + \mathbf{b}u \\ y - \tilde{y} = \mathbf{c}^\top \mathbf{x} \end{cases}$$

has a strictly positive real transfer function, $H(s)$, and contains an unknown parameters vector, θ, where $\tilde{y} = \tilde{y}(t)$ is a given target trajectory for tracking. Let $\widehat{\theta} = \widehat{\theta}(t)$ be an estimate of θ, generated by the following observer, with

$$e_y := y - \tilde{y}.$$

If, in the frequency domain after taking the Laplace transformations,

$$E_y(s) = \mathcal{L}\{e_y(t)\} = H(s)\mathcal{L}\left\{\alpha\,\widehat{\theta}^\top \mathbf{v}(t)\right\}$$

for some vector-valued function $\mathbf{v}(t)$ and an unknown constant α with known (positive or negative) sign, then using an observer for estimating θ, in the form of

$$\dot{\widehat{\theta}} = -\text{sign}\{\alpha\}\,\kappa\,e_y(t)\mathbf{v}(t),$$

where $\kappa > 0$ is a constant and $\widehat{\theta}$ is bounded, the controller

$$u(t) = \alpha\,\widehat{\theta}^\top \mathbf{v}(t)$$

will yield a globally bounded error trajectory, $e_y(t)$. If, furthermore, $\|\mathbf{v}\| \le M < \infty$, then

$$e_y(t) \to 0 \quad \text{namely} \quad y(t) \to \tilde{y}(t) \quad (t \to \infty).$$

Proof. The controlled system is

$$\begin{cases} \dot{\mathbf{x}} = A\mathbf{x} + \mathbf{b}\left(\alpha\,\widehat{\theta}^\top \mathbf{v}(t)\right) \\ e_y = \mathbf{c}^\top \mathbf{x}. \end{cases}$$

It follows from Lemma 6.1 that, since $H(s)$ is strictly positive real, there exist positive definite and symmetrical matrices P and Q such that

$$\begin{cases} A^\top P + PA = Q \\ P\mathbf{b} = \mathbf{c}. \end{cases}$$

Construct a Lyapunov function

$$V(\mathbf{x}, \widehat{\theta}) = \mathbf{x}^\top P\mathbf{x} + \frac{|\alpha|}{\kappa}\,\widehat{\theta}^\top \widehat{\theta}.$$

Then, one has

$$\dot{V} = \mathbf{x}^\top P\dot{\mathbf{x}} + \dot{\mathbf{x}}^\top P\mathbf{x} + \tfrac{|\alpha|}{\kappa}\left(\left(\dot{\widehat{\theta}}\right)^\top\widehat{\theta} + \left(\widehat{\theta}\right)^\top\dot{\widehat{\theta}}\right)$$

$$= \mathbf{x}^\top P\left(A\mathbf{x} + \mathbf{b}\,\alpha\,\widehat{\theta}^\top\mathbf{v}(t)\right) + \left(A\mathbf{x} + \mathbf{b}\,\alpha\,\widehat{\theta}^\top\mathbf{v}(t)\right)^\top P\mathbf{x}$$

$$+ \tfrac{|\alpha|}{\kappa}\left(-2\,\text{sign}\{\alpha\}\,\kappa\,\mathbf{v}^\top\widehat{\theta}\right)$$

$$= \mathbf{x}^\top(PA + A^\top P)\mathbf{x} + 2\,\mathbf{x}^\top Pb\alpha\widehat{\theta}^\top\mathbf{v} - 2\widehat{\theta}^\top\alpha\,e_y\,\mathbf{v}$$

$$= \mathbf{x}^\top(-Q)\mathbf{x} + 2\,\mathbf{x}^\top\mathbf{c}\alpha\widehat{\theta}^\top\mathbf{v} - 2\widehat{\theta}^\top\alpha\,\mathbf{c}^\top\mathbf{x}\,\mathbf{v}$$

$$= -\mathbf{x}^\top Q\mathbf{x}$$

$$< 0 \qquad \text{(for all } \mathbf{x} \neq 0\text{)}.$$

Therefore, the system defined by

$$\begin{cases} \dot{\mathbf{x}} = A\mathbf{x} + \mathbf{b}\left(\alpha\,\widehat{\theta}^\top\mathbf{v}(t)\right) \\ \dot{\widehat{\theta}} = -\text{sign}\{\alpha\}\,\kappa\,\mathbf{c}^\top\mathbf{x}\mathbf{v}(t) \end{cases}$$

is globally asymptotically stable about its zero fixed point $(\mathbf{x}^*, \widehat{\theta}^*)$. If, furthermore, $||\mathbf{v}|| \leq M < \infty$, then

$$e_y = \mathbf{c}^\top\mathbf{x} \to 0 \qquad (t \to \infty),$$

as claimed.   □

6.3.2.3 *Adaptive Control of Uncertain Nonlinear Systems*

The above methodology can also be extended to some uncertain nonlinear systems, as shown by the following simple example.

Example 6.5. Consider an uncertain nonlinear system,

$$\dot{x} + \alpha\,g(x) = u, \qquad g(0) = 0,$$

where α is an unknown constant and $g(\cdot)$ is a given nonlinear function. The objective is to design a controller, $u = u(t)$, to derive $x(t) \to 0$ as $t \to \infty$.

To have some mathematical idea about the controller design in the case that α is known, one may attempt to use the controller

$$u = \alpha\,g(x) - cx \qquad (c > 0),$$

which leads to $\dot{x} + cx = 0$ with a solution $x(t) = x_0 e^{-ct} \to 0$ as $t \to \infty$. This controller is clearly impractical since it simply cancels the nonlinear term, but it offers some idea for a practical design as further discussed below.

To follow the above naive idea, but now α is unknown, so a simple controller seems impossible. In this situation, perhaps the simplest but acceptable choice could be to design an adaptive controller of the form

$$u = \widehat{\alpha}\, g(x) - cx \qquad (c > 0),$$

where $\widehat{\alpha}$ is an estimate of the unknown α. Then, the controlled system is

$$\dot{x} + e_\alpha\, g(x) + cx = 0\,,$$

where $e_\alpha = \alpha - \widehat{\alpha}$ is the estimation error.

Try the Lyapunov function

$$V(x, e_\alpha) = \frac{1}{2}x^2 + \frac{1}{2\beta}\, e_\alpha^2 \qquad (\beta > 0),$$

which gives

$$
\begin{aligned}
\dot{V} &= x\dot{x} + \tfrac{1}{\beta}\, e_\alpha \dot{e}_\alpha \\
&= x(-cx - e_\alpha g(x)) + \tfrac{1}{\beta}\, e_\alpha \dot{e}_\alpha \\
&= -cx^2 - e_\alpha\left(xg(x) - \tfrac{1}{\beta}\dot{e}_\alpha\right).
\end{aligned}
$$

Hence, one may design an observer to estimate α, such that

$$\dot{\widehat{\alpha}} = -\beta x g(x)\,,$$

which yields

$$\dot{e}_\alpha = -\dot{\widehat{\alpha}} = \beta x g(x)\,,$$

so that

$$\dot{V} = -cx^2 < 0 \qquad \text{for all } x \neq 0\,.$$

Thus, the controller is obtained as

$$u = \widehat{\alpha}g(x) - cx \qquad (c > 0)\,,$$

where

$$\widehat{\alpha}(t) = \widehat{\alpha}(0) - \beta \int_0^t x(\tau)g(x(\tau))d\tau\,,$$

with $\widehat{\alpha}(0) = 0$ and an arbitrary $\beta > 0$.

6.3.3 *Lyapunov Redesign of Nonlinear Controllers*

Consider a nominal model of nonlinear control systems,

$$\dot{\mathbf{x}} = \mathbf{f}(\mathbf{x}, t) + G(\mathbf{x}, t)\mathbf{u}, \tag{6.32}$$

where $\mathbf{f}$ and G are nonlinear vector- or matrix-valued functions satisfying $\mathbf{f}(0, t) = 0$ and $G(0, t) = 0$ for all $t \geq t_0$, with a state-feedback controller

$$\mathbf{u} = \mathbf{u}_1(\mathbf{x}, t)$$

to be designed.

Suppose that the controller stabilizes the nominal system, such that the controlled system

$$\dot{\mathbf{x}} = \mathbf{f}(\mathbf{x}, t) + G(\mathbf{x}, t)\mathbf{u}_1(\mathbf{x}, t)$$

is uniformly asymptotically stable about its zero fixed point.

In this case, there exists a Lyapunov function, $V(\mathbf{x}, t)$, and three class-$\mathcal{K}$ functions, $\alpha(\cdot)$, $\beta(\cdot)$ and $\gamma(\cdot)$, such that (see Sec. 2.2)

$$\alpha(||\mathbf{x}||) \leq V(\mathbf{x}) \leq \beta(||\mathbf{x}||)$$

and

$$\dot{V}(\mathbf{x}, t) \leq -\gamma(||\mathbf{x}||)$$

for all $\mathbf{x} \in \mathcal{D}$, the defining domain of the system, which contains the origin.

Now, suppose that the nominal system is perturbed by some uncertainty, in the form of an external input, leading to

$$\dot{\mathbf{x}} = \mathbf{f}(\mathbf{x}, t) + G(\mathbf{x}, t)[\mathbf{u} + \delta(\mathbf{x}, \mathbf{u}, t)], \tag{6.33}$$

where δ is an unknown vector-valued function, which is small such that the perturbed system (6.33) remains uniquely solvable under the original initial conditions. Further, suppose that, for any control input $\mathbf{u} = \mathbf{u}_1 + \mathbf{u}_2$, with $\mathbf{u}_2$ satisfying

$$||\delta((\mathbf{x}, \mathbf{u}_1 + \mathbf{u}_2, t)|| \leq h(\mathbf{x}, t) + c||\mathbf{u}_2||$$

for a non-negative continuous function $h(\cdot, t)$, which need not be small in magnitude, with a constant $0 \leq c < 1$.

Then, design $\mathbf{u}_2$ such that the controller $\mathbf{u} = \mathbf{u}_1 + \mathbf{u}_2$ can stabilize the perturbed system (6.33). This design of the additional controller $\mathbf{u}_2$ is referred to as a *redesign* of the first design of the controller $\mathbf{u} = \mathbf{u}_1$.

To complete the redesign, start with

$$\begin{aligned}
\dot{\mathbf{x}} &= \mathbf{f}(\mathbf{x}, t) + G(\mathbf{x}, t)[\mathbf{u} + \delta(\mathbf{x}, \mathbf{u}, t)] \\
&= \mathbf{f}(\mathbf{x}, t) + G(\mathbf{x}, t)\mathbf{u}_1 + G(\mathbf{x}, t)[\mathbf{u}_2 + \delta(\mathbf{x}, \mathbf{u}_1 + \mathbf{u}_2, t)].
\end{aligned}$$

If the perturbed system (6.33) can be stabilized then there exists a Lyapunov function, $\widetilde{V}(\mathbf{x}, t)$, associated with the perturbed system, which reduces to the Lyapunov function $V(\mathbf{x}, t)$ for the nominal system when $\delta = 0$. This Lyapunov function $\widetilde{V}(\mathbf{x}, t)$ satisfies

$$\alpha(||\mathbf{x}||) \leq \widetilde{V}(\mathbf{x}) \leq \beta(||\mathbf{x}||),$$

for some class-$\mathcal{K}$ functions α and β that satisfy both of the above two inequalities.

It follows that

$$\begin{aligned}
\dot{\widetilde{V}} &= \tfrac{\partial \widetilde{V}}{\partial t} + \tfrac{\partial \widetilde{V}}{\partial \mathbf{x}}(\mathbf{f} + G\mathbf{u}_1) + \tfrac{\partial \widetilde{V}}{\partial \mathbf{x}}(\mathbf{u}_2 + \delta) \\
&= \tfrac{\partial V}{\partial t} + \tfrac{\partial V}{\partial \mathbf{x}}(\mathbf{f} + G\mathbf{u}_1) + \tfrac{\partial \widetilde{V}}{\partial \mathbf{x}}(\mathbf{u}_2 + \delta) \\
&\leq -\gamma(||\mathbf{x}||) + \tfrac{\partial \widetilde{V}}{\partial \mathbf{x}}(\mathbf{u}_2 + \delta)
\end{aligned}$$

for all $\mathbf{x} \in \mathcal{D}$. Denote $\mathbf{w}^\top = [\partial \widetilde{V}/\partial \mathbf{x}]G$ and observe that

$$\begin{aligned}
\mathbf{w}^\top(\mathbf{u}_2 + \delta) &= \mathbf{w}^\top \mathbf{u}_2 + \mathbf{w}^\top \delta \\
&\leq \mathbf{w}^\top \mathbf{u}_2 + ||\mathbf{w}|| \cdot ||\delta|| \\
&\leq \mathbf{w}^\top \mathbf{u}_2 + ||\mathbf{w}|| \cdot [h(\mathbf{x}, t) + c||\mathbf{u}_2||].
\end{aligned}$$

Then, choose

$$\mathbf{u}_2 \leq -\frac{\rho(\mathbf{x}, t)}{1 - c} \cdot \frac{\mathbf{w}}{||\mathbf{w}||},$$

where $h(\mathbf{x}, t) \leq \rho(\mathbf{x}, t)$ for all $\mathbf{x} \in \mathcal{D}$ component-wise for all $t \geq t_0$. It follows that

$$\mathbf{w}^\top(\mathbf{u}_2 + \delta) \leq \frac{\rho(\mathbf{x}, t)}{1 - c}||\mathbf{w}|| + ||\mathbf{w}||\rho(\mathbf{x}, t) + \frac{c}{1 - c}||\mathbf{w}||\rho(\mathbf{x}, t) \leq 0,$$

implying that

$$\dot{\widetilde{V}}(\mathbf{x}, t) \leq -\gamma(||\mathbf{x}||).$$

This guarantees the uniform asymptotic stability of the perturbed system (6.33) about its zero fixed point.

Example 6.6. Consider the nonlinear controlled pendulum equation

$$\begin{cases} \dot{x} = y \\ \dot{y} = -\tfrac{\kappa}{m} y - a \sin(x) + b y \end{cases}$$

with $a = g/\ell$, where the gravity g, the length ℓ and the coefficient b are all constants.

Following the above discussions to design a controller that can stabilize this pendulum system, try

$$u_1 = b\,u = a\,\sin(x) + k_1 x + k_2 y$$

with k_1 and k_2 being chosen such that the matrix $\begin{bmatrix} 0 & 1 \\ k_1 & k_2 - \kappa/m \end{bmatrix}$ is stable.

Now, suppose that the pendulum system contains some uncertainties, so that the constants a and b are unknown hence cannot be used for the controller.

Let the perturbed parameters be $\tilde{a}$ and $\tilde{b}$. Then, it can be verified that the above perturbation term δ is given by

$$\delta(\mathbf{x}, t) = \frac{a\tilde{b} - \tilde{a}b}{b}\,\sin(x) + \frac{\tilde{b} - b}{b}(u_1 - a\,\sin(x)),$$

where $\mathbf{x} = [x\ y]^{\top}$. Therefore,

$$\|\delta(\mathbf{x}, u_1 + u_2)\|$$
$$\leq \left|\frac{a\tilde{b} - \tilde{a}b}{b}\right| \|x\| + \left|\frac{\tilde{b} - b}{b}\right| \|u_1 - a\,\sin(x)\| + \left|\frac{\tilde{b} - b}{b}\right| \|u_2\|$$
$$\leq \left|\frac{a\tilde{b} - \tilde{a}b}{b}\right| \|x\| + \left|\frac{\tilde{b} - b}{b}\right| |k_1| \|x\| + \left|\frac{\tilde{b} - b}{b}\right| |k_2| \|y\| + \left|\frac{\tilde{b} - b}{b}\right| \|u_2\|$$
$$:= h(\mathbf{x}) + c\|u_2\|,$$

where $h(\cdot)$ is a non-negative continuous function.

It then follows from the Lyapunov redesign method that the new feedback controller, u_2, needs to be found by constructing a suitable Lyapunov function. To do so, let P be a solution of the following matrix Lyapunov equation:

$$P \begin{bmatrix} 0 & 1 \\ k_1 & k_2 - \kappa/m \end{bmatrix} + P \begin{bmatrix} 0 & k_1 \\ 1_1 & k_2 - \kappa/m \end{bmatrix} P + I = 0$$

and calculate the scaler-valued function $w(\mathbf{x})$ by

$$w(\mathbf{x}) = \frac{\partial \tilde{V}}{\partial \mathbf{x}} G = \frac{\partial}{\partial \mathbf{x}}\left\{\frac{1}{2}\mathbf{x}^{\top}P\mathbf{x}\right\} \begin{bmatrix} 0 \\ b \end{bmatrix} = \mathbf{x}^{\top} P \begin{bmatrix} 0 \\ b \end{bmatrix}.$$

It then follows that

$$u_2 \leq -\frac{\rho(\mathbf{x})}{1 - c}\frac{w(\mathbf{x})}{|w(\mathbf{x})|},$$

where $\rho(\mathbf{x}) \geq h(\mathbf{x})$ component-wise, which can be any non-negative continuous function providing some freedom for other purposes such as robust stability of the controlled system.

6.3.4 *Sliding-Mode Control*

Sliding-mode control, also known as variable structure control, is a nonlinear control method by which the motion dynamics of the controlled system is effectively constrained to be within a subspace of the system state space. The sliding motion is then achieved by a controller or algorithm to alter the system dynamics along some consequent sliding-mode surfaces within the state space, where on each sliding mode surface the system is equivalent to a lower-order uncontrolled system that is easier to control.

Sliding-mode control is achieved in two steps: (i) (sliding-mode surface design) predefine a sliding surface, in such a way that the desired system dynamics can be achieved during the sliding-mode motion; (ii) (sliding-mode controller design) design a controller to drive the closed-loop system dynamics to reach and then be retained on the sliding surface. These are respectively discussed in the following.

6.3.4.1 *Sliding-Mode Surface Design*

To introduce the main idea of sliding-mode control, consider an LTI system,

$$\dot{\mathbf{x}} = A\mathbf{x} + B\mathbf{u}, \tag{6.34}$$

where $\mathbf{x} \in R^n$ and $\mathbf{u} \in R^m$ with $1 \le m \le n$. Assume that this LIT system is controllable.

Predefine a variable vector $\mathbf{s}(t) \in R^m$, which passes the origin of the state space, in the following form

$$\mathbf{s}(t) = C\mathbf{x}(t), \tag{6.35}$$

where $C \in R^{m \times n}$ is a parametric matrix such that CB is nonsingular, which is to be further determined.

Introduce an *equivalent controller*, $\mathbf{u}_{eq} \in R^m$, such that it can restrict the system motion on the sliding-mode surface $\mathbf{s}(t) = 0$ with $\dot{\mathbf{s}}(t) = 0$ for all $t \ge t_0$. Thus,

$$\begin{cases} \dot{\mathbf{s}}(t) = C\dot{\mathbf{x}}(t) = 0 \\ C\big[A\mathbf{x}(t) + B\mathbf{u}_{eq}(t)\big] = 0, \end{cases} \tag{6.36}$$

yielding

$$\mathbf{u}_{eq} = -[CB]^{-1}CA\mathbf{x} \tag{6.37}$$

and

$$\dot{\mathbf{x}}(t) = \big[I - B[CB]^{-1}C\big]A\mathbf{x}(t). \tag{6.38}$$

The system (6.34) is said to remain on the sliding-mode surface if its state vector reaches the surface and then sliding on all the connected switching surfaces in the state space.

Figure 6.6 visualizes the "reaching" phase and "sliding" phase of the sliding-mode control in the 1-dimensional system setting.

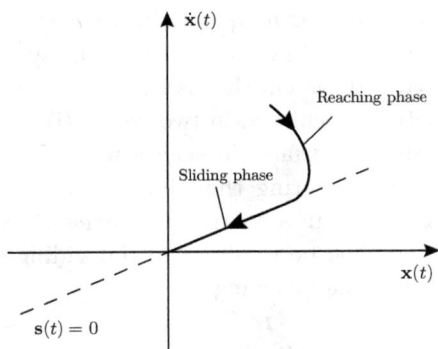

Fig. 6.6   Sliding-model control: "reaching" phase and "sliding" phase.

In the reaching phase of the sliding-mode controller, the Lyapunov second method described by Theorem 2.2 is useful. Consider the Lyapunov function

$$V = \frac{1}{2}\mathbf{s}^{\top}\mathbf{s},$$

which satisfies

$$\dot{V} = \mathbf{s}^{\top}(t)\dot{\mathbf{s}}(t) < 0, \qquad \text{for } \mathbf{s}(t) \neq 0, \tag{6.39}$$

if the *reaching condition*

$$\mathbf{s}^{\top}(t)\dot{\mathbf{s}}(t) < 0 \tag{6.40}$$

is satisfied.

6.3.4.2   *Sliding-Mode Controller Design*

Under the (sufficient) condition (6.40), a *reaching law* is designed as

$$\dot{\mathbf{s}}(t) = -\varepsilon \, \text{sgn}[\mathbf{s}(t)] - \mathbf{g}(\mathbf{x}(t)), \tag{6.41}$$

where the signum function $\text{sgn}[\cdot]$ is defined component-wise, with a constant control gain $\varepsilon > 0$ and continuous function $\mathbf{g}$ satisfying $\mathbf{g}(0) = 0$ and $\mathbf{s}^{\top}\mathbf{g}(\mathbf{s}(t)) > 0$ for $\mathbf{s}(t) \neq 0$ and all $t \geq t_0$.

Example 6.7. Some specific reaching laws (6.41) are:

(i) $\dot{\mathbf{s}}(t) = -\varepsilon \operatorname{sgn}[\mathbf{s}(t)]$;
(ii) $\dot{\mathbf{s}}(t) = -\varepsilon \operatorname{sgn}[\mathbf{s}(t)] - k\mathbf{s}(t)$, $k > 0$;
(iii) $\dot{\mathbf{s}}(t) = -\varepsilon \, ||\mathbf{s}(t)||^k \operatorname{sgn}[\mathbf{s}(t)]$, $0 < k < 1$.

A typical sliding-mode controller is composed of two parts:

$$\mathbf{u}(t) = \mathbf{u}_{eq}(t) + \mathbf{u}_s(t), \qquad (6.42)$$

where $\mathbf{u}_{eq}(t)$ is the equivalent controller (6.37) and

$$\mathbf{u}_s(t) = -h[CB]^{-1}\operatorname{sgn}[\mathbf{s}(t)] \qquad (6.43)$$

is a simple nonlinear controller with $h > 0$ being a constant control gain to be determined.

Substituting (6.42), namely both (6.37) and (6.43), into (6.39) yields

$$\dot{V}(t) = \mathbf{s}^\top(t)[CA]\mathbf{x}(t) + \mathbf{s}^\top[CB]\mathbf{u}(t)$$
$$= -h||\mathbf{s}(t)|| < 0.$$

Consequently, $\mathbf{s}(t) \to 0$ as $t \to \infty$.

After the system state vector reaches the sliding-mode surface $\mathbf{s}(t) = 0$, the system dynamics will stay in the sliding phase governed by (6.38).

6.3.4.3 *Chattering and its Attenuation*

During the sliding phase, due to possible switching time delay in the sign change of $\mathbf{s}(t)$ and switch device time-lag in implementation, the system trajectory may change its motion direction from time to time, or even frequently. Such repetition of direction changes generates a "zig-zag motion" oscillating around the predefined sliding surface, as illustrated in Fig. 6.7.

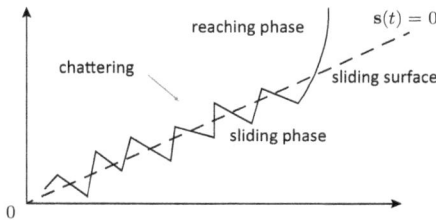

Fig. 6.7   Chattering in the sliding-mode control process.

In general, the chattering will result in low control accuracy, non-smooth motion performance, high heat losses in circuits, and high wear of mechanical components, and even instability; therefore, it is very undesirable. In the worst situation, at least in theory, there could even be the so-called "Zeno phenomenon" in the sense that infinitely many times of switchings occur in a finite time duration. All such chattering oscillations have to be reduced or, if possible, completely eliminated.

One simple but effective technique to handle the chattering problem is to use the so-called *boundary-layer* controller, namely, to replace the second part of the sliding-mode controller (6.43) by

$$\mathbf{u}_s(t) = -h[CB]^{-1} \operatorname{sat}[\mathbf{s}(t)], \tag{6.44}$$

with the saturation function

$$\operatorname{sat}[\mathbf{s}(t)] = \begin{cases} \mathbf{s}(t)/\rho & \text{for } ||\mathbf{s}(t)|| \leq \rho \\ \operatorname{sgn}[\mathbf{s}(t)] & \text{for } ||\mathbf{s}(t)|| > \rho, \end{cases}$$

where the constant $\rho > 0$ is chosen to well control the boundaries of the chattering motion.

It should be noted that, although the boundary-layer controller may reduce or even eliminate the chattering motion, it may lead to losing the system asymptotic stability, therefore this is not a perfect solution. As a remedy or improvement, the following continuous approximation can be considered:

$$\mathbf{u}_s(t) = -h[CB]^{-1} \left(\frac{\mathbf{s}(t)}{||\mathbf{s}(t)|| + \rho} \right), \tag{6.45}$$

However, this modification is not perfect either, since it needs a high-gain control when the system trajectory is close to the sliding surface.

Therefore, more advanced techniques are needed to perform better control, which has seen promising progress, but will not be further discussed in this textbook.

6.4   Controlling Bifurcations and Chaos

6.4.1   *Controlling Bifurcations*

The notion of bifurcations in nonlinear dynamical systems, studied in Sec. 5.1, is important and useful for many engineering applications. Various types of bifurcations occur frequently in practical systems, which are usually harmful to the systems regarding their structures and dynamics.

Therefore, they should be greatly reduced or even completely eliminated, which is a special task commonly referred to as *bifurcation control*.

Typical examples of bifurcation control include crisis monitoring and protection of power grids, dynamics-based practical design of power electronic devices, surge and stall prevention for jet engines, regulation of pathological heart rhythms (e.g. fibrillation, ectopic foci), electronic and mechanical control systems analysis and design such as thermal convection, lasers, oscillators, and so on.

This section is devoted to a study of the bifurcation control task, presenting some basic ideas, design principles and analysis methods.

6.4.1.1 *Controlling Discrete-Time Nonlinear Systems*

In applications, sometimes it is important to delay the onset of a bifurcation of stability to gain some time for taking actions to protect a system. For instance, for the logistic map (system), one desirable but challenging task is to delay the appearance of period-doubling bifurcation or the onset of chaos as its key parameter approaches the bifurcation value, by implementing a simple (e.g. linear) state-feedback controller without directly modifying the parameter, which is usually unaccessible as discussed in Sec. 6.1.1.

As an example, consider the logistic map (5.3), namely

$$x_{k+1} = f(x_k, p) := p\, x_k\, (1 - x_k).$$

As the parameter p is gradually increased in interval $(3.0, 4.0)$, it shows period-doubling leading to chaos:

$$\text{period}: \ 1 \to 2 \to 4 \to \cdots \to 2^k \to \cdots \to \text{chaos}$$

as shown in Fig. 5.9.

Now, consider the bifurcation control problem as follows:

Is it possible (and if so, how) to design a simple controller (a control sequence), $\{u_k\}$, such that the controlled logistic system

$$x_{k+1} = F(x_k, p) := p\, x_k(1 - x_k) + u_k$$

can achieve, for example, one of the following goals:

(i) the limiting chaotic behavior of the period-doubling bifurcation process is delayed, or completely suppressed;

(ii) the first or second bifurcation has delayed occurrence, or some bifurcations are changed either in form or in stability;

(iii) the asymptotic behavior of the controlled system becomes chaotic, if desirable, when the parameter p is currently not in the chaotic region.

In this section, for some typical bifurcations, the problems of controlling transcritical, saddle-node, pitchfork, Hopf and period-doubling bifurcations are studied.

To prepare for these, their basic properties are first reviewed (see [Glendinning (1994)]).

(a) Transcritical Bifurcation

Theorem 6.4.   *Consider the following system:*

$$\dot{x} = f(x; p), \quad f(0,0) = \frac{\partial f}{\partial x}(0,0) = 0.$$

If

$$\frac{\partial f}{\partial p}(0,0) = 0, \quad \frac{\partial^2 f}{\partial x^2}(0,0) \neq 0$$

$$\frac{\partial^2 f}{\partial p \partial x}(0,0) - \frac{\partial^2 f}{\partial x^2} \cdot \frac{\partial^2 f}{\partial p^2}(0,0) > 0,$$

then the system has a transcritical bifurcation at $(0,0)$. Moreover,

$\frac{\partial^2 f}{\partial x^2}(0,0) > 0 \;\rightarrow\;$ *stable/unstable branches*

$\frac{\partial^2 f}{\partial x^2}(0,0) < 0 \;\rightarrow\;$ *unstable/stable branches*

(b) Saddle-Node Bifurcation

Theorem 6.5.   *Consider the following system:*

$$\dot{x} = f(x; p), \quad f(0,0) = \frac{\partial f}{\partial x}(0,0) = 0.$$

If

$$\frac{\partial f}{\partial p}(0,0) \neq 0, \quad \frac{\partial^2 f}{\partial x^2}(0,0) \neq 0,$$

then the system has a saddle-note bifurcation at $(0,0)$. Moreover,

$\frac{\partial^2 f}{\partial x^2}(0,0) > 0 \;\rightarrow\;$ *stable/unstable branches*

$\frac{\partial^2 f}{\partial x^2}(0,0) < 0 \;\rightarrow\;$ *unstable/stable branches*

(c) Pitchfork Bifurcation

Theorem 6.6. *Consider the following system:*

$$\dot{x} = f(x; p), \quad f(0,0) = \frac{\partial f}{\partial x}(0,0) = 0.$$

If

$$\frac{\partial f}{\partial p}(0,0) = \frac{\partial^2 f}{\partial x^2}(0,0) = 0$$

$$\frac{\partial^2 f}{\partial p \partial x}(0,0) \neq 0, \quad \frac{\partial^3 f}{\partial x^3}(0,0) \neq 0,$$

then the system has a pitchfork bifurcation at $(0,0)$. *Moreover,*

$\frac{\partial^3 f}{\partial x^3}(0,0) > 0 \quad \rightarrow \quad$ *stable/unstable branches*

$\frac{\partial^3 f}{\partial x^3}(0,0) < 0 \quad \rightarrow \quad$ *unstable/stable branches*

(d) Hopf Bifurcation

Recall the Hopf Bifurcation Theorem 5.3, as follows.

Theorem 6.7. *Suppose that the 2-dimensional system*

$$\dot{x} = f(x, y; p)$$
$$\dot{y} = g(x, y; p)$$

has a zero equilibrium, $(x^*, y^*) = (0,0)$, *and that its Jacobian has a pair of complex conjugate eigenvalues,* $\lambda(p)$ *and* $\bar{\lambda}(p)$. *If, for some* p_0,

$$\frac{d\,\text{Re}\{\lambda(p)\}}{d\,p}\bigg|_{p=p_0} > 0,$$

then $p = p_0$ *is a bifurcation point, and*

(i) *for close enough values* $p < p_0$, *the zero equilibrium is asymptotically stable;*

(ii) *for close enough values* $p > p_0$, *the zero equilibrium is unstable;*

(iii) *for close enough values* $p \neq p_0$, *the zero equilibrium is surrounded by a limit cycle of magnitude* $O(\sqrt{|p - p_0|})$.

The discrete version of the Hopf bifurcation theorem is introduced as follows.

Consider a 2-dimensional parameterized system:

$$x_{k+1} = f(x_k, y_k; p),$$
$$y_{k+1} = g(x_k, y_k; p),$$

with a parameter p and an equilibrium point (x^*, y^*), satisfying $x^* = f(x^*, y^*; p)$, $y^* = g(x^*, y^*; p)$ simultaneously for all p.

Let $J(p)$ be its Jacobian at this equilibrium, with a pair of complex conjugate eigenvalues, $\lambda_2(p) = \bar{\lambda}_1(p)$. If

$$|\lambda_1(p^*)| = 1 \quad \text{and} \quad \left. \frac{\partial |\lambda_1(p)|}{\partial p} \right|_{p=p^*} > 0$$

then the system has a Hopf bifurcation at (x^*, y^*, p^*), in a way analogous to the continuous-time setting.

6.4.1.2   *State-Feedback Control of Bifurcations*

Consider a 1-dimensional discrete control system,

$$x_{k+1} = F(x_k; p) := f(x_k; p) + u(x_k; p),$$

where p is a real parameter, x_0 is the initial state, and $u(\cdot)$ is a state-feedback controller to be designed.

Usually, $u(\cdot; p)$ is required not to modify the system equilibria, e.g. $u(0; p) = 0$, in order not to alter the original system dynamics when control is nonexistent.

Bifurcation controller design for this system is formulated as the following routine-checking procedure.

Design Scheme

Step 0. Initiate a structure of the controller (e.g. a linear state-feedback controller):

$$x_{k+1} = F(x_k; p) := f(x_k; p) + u(x_k; p).$$

Step 1. Solve the equilibrium equation

$$x^* = F(x^*; p)$$

for a solution $x^*(t; p)$; if no solution, change the structure of the controller and try again.

Table 6.1   Classification of bifurcations.

$\frac{\partial F}{\partial p}\big|_{(x^*,p^*)}$	$\frac{\partial^2 F}{\partial x^2}\big|_{(x^*,p^*)}$	Bifurcations
$\neq 0$	$\neq 0$	saddle-node
$= 0$	$\neq 0$	transcriptical
$= 0$	$= 0$	pitchfork

Table 6.2   Stability of equilibria near saddle-node bifurcation.

$\frac{\partial F}{\partial p}\big|_{(x^*,p^*)}$	$\frac{\partial^2 F}{\partial x^2}\big|_{(x^*,p^*)}$	stable equil.	unstable equil.	no equil.
> 0	> 0	$p < p^*$ (upper)	$p < p^*$ (lower)	$p > p^*$
> 0	< 0	$p > p^*$ (upper)	$p > p^*$ (lower)	$p < p^*$
< 0	> 0	$p > p^*$ (upper)	$p > p^*$ (lower)	$p < p^*$
< 0	< 0	$p < p^*$ (upper)	$p < p^*$ (lower)	$p > p^*$

Table 6.3   Stability of equilibria near transcritical bifurcation.

$\frac{\partial^2 F}{\partial x \partial p}\big|_{(x^*,p^*)} - \frac{\partial^2 F}{\partial x^2}\frac{\partial^2 F}{\partial p^2}$	$\frac{\partial^2 F}{\partial x^2}\big|_{(x^*,p^*)}$	stable equil.	unstable equil.
$\neq 0$	> 0	(lower)	(upper)
$\neq 0$	< 0	(upper	(lower)

Step 2. Determine the bifurcating parameter value $p = p^*$ such that
$$\frac{\partial F}{\partial x}\bigg|_{\substack{x=x^* \\ p=p^*}} = 1\,;$$
if no solution, change the structure of the controller, and then return to Step 1.

Step 3. Determine the type of the bifurcation according to Table 6.1.

Step 4. Determine the stability of the equilibria according to Tables 6.2–6.4.

    Several controller designs for some typical 1-dimensional system bifurcations are specified below.

Table 6.4   Stability of equilibria near pitchfork bifurcation.

$\frac{\partial F}{\partial p}\Big|_{(x^*,p^*)}$	$\frac{\partial^3 F}{\partial x^3}\Big|_{(x^*,p^*)}$	stable equil. (1st branch)	unstable equil. (1st branch)	stable equil. (2nd branch)	unstable equil. (2nd branch)
> 0	> 0	$p < p^*$	$p > p^*$	—	$p < p^*$
> 0	< 0	$p < p^*$	$p > p^*$	$p > p^*$	—
< 0	> 0	$p > p^*$	$p < p^*$	—	$p > p^*$
< 0	< 0	$p > p^*$	$p < p^*$	$p < p^*$	—

Table 6.5   Stability near period-doubling bifurcation.

ξ	η	period-doubling	stable equil.	unstable equil.
> 0	> 0	$p < p^*$ (stable)	$p > p^*$	$p < p^*$
> 0	< 0	$p > p^*$ (unstable)	$p > p^*$	$p < p^*$
< 0	> 0	$p > p^*$ (stable)	$p < p^*$	$p > p^*$
< 0	< 0	$p < p^*$ (unstable)	$p < p^*$	$p > p^*$

6.4.1.3   *Controlling Period-Doubling Bifurcation*

Design Scheme

Step 0. Initiate a structure of the controller.

Step 1. Solve the equilibrium equation $x^* = F(x^*; p)$ for a solution $x^*(t; p)$; if no solution, change the structure of the controller and try again.

Step 2. Determine the bifurcating parameter $p = p^*$ such that

$$\frac{\partial F}{\partial x}\bigg|_{\substack{x=x^* \\ p=p^*}} = -1\,;$$

if no solution, change the controller structure and then return to Step 1.

Step 3. Determine the existence of the period-doubling bifurcation and the stability of the equilibria according to Table 6.5, in which

$$\xi = \left(2\,\frac{\partial^2 F}{\partial x \partial p} + \frac{\partial F}{\partial p}\,\frac{\partial^2 F}{\partial x^2} \right)\bigg|_{\substack{x=x^* \\ p=p^*}}\,,$$

$$\eta = \left(\frac{1}{2}\left(\frac{\partial^2 F}{\partial x^2} \right)^2 + \frac{1}{3}\,\frac{\partial^3 F^2}{\partial x^3} \right)\bigg|_{\substack{x=x^* \\ p=p^*}}\,.$$

Example 6.8. Consider the controlled logistic system:
$$x_{k+1} = F(x_k, p) = p\,x_k(1 - x_k) + u_k\,.$$
The objective is to shift the original bifurcation point (x^*, p^*) to a new position, (x^o, p^o).

Following the above procedure, solving the equilibrium equation
$$F(x^o, p^o) = p^o\,x^o(1 - x^o) + u\Big|_{\substack{x=x^o \\ p=p^o}} = x^o$$

gives
$$u\Big|_{\substack{x=x^o \\ p=p^o}} = x^o - p^o x^o + p^o (x^o)^2\,.$$

Then, the control condition leads to
$$\frac{\partial F}{\partial x}\Big|_{\substack{x=x^o \\ p=p^o}} = p^o - 2p^o x^o + \frac{\partial u}{\partial x}\Big|_{\substack{x=x^o \\ p=p^o}} = 1\,,$$

which gives
$$\frac{\partial u}{\partial x}\Big|_{\substack{x=x^o \\ p=p^o}} = 1 - p^o + 2p^o x^o\,.$$

These two conditions together yield a simple linear controller:
$$u(x_k; p) = \left(1 - p^o + 2p^o x^o\right) x_k - p^o (x^o)^2 =: c_1 x_k + c_2\,.$$

Thus, for the linearly controlled logistic map
$$x_{k+1} = F(x_k; p) = p\,x_k(1 - x_k) + (c_1 x_k + c_2)\,,$$

one has
$$\frac{\partial F}{\partial p}\Big|_{\substack{x=x^o \\ p=p^o}} = x^o - (x^o)^2\,, \qquad \frac{\partial^2 F}{\partial x^2}\Big|_{\substack{x=x^o \\ p=p^o}} = -2p^o\,.$$

According to Table 6.1:

- if $x^o = 0$ and $p^o \neq 0$, then (x^o, p^o) is a transcritical bifurcating point;
- if $x^o \neq 0$ and $p^o \neq 0$, then (x^o, p^o) is a saddle-node bifurcating point;
- if $x^o = 0$ and $p^o = 0$, then (x^o, p^o) is a pitchfork bifurcating point.

Note that for pitchfork bifurcation, Table 6.4 shows that $\partial^3 F/\partial x^3 \neq 0$, so a linear controller cannot produce a pitchfork bifurcation. However, actually an additional cubic term can easily do the job, though the controller is more complex.

On the other hand, one has
$$\xi = 2(1 - 2x^o) - 2p^o x^o(1 - x^o), \qquad \eta = 2(p^o)^2\,.$$
Thus, a controller can be designed to shift the original period-doubling bifurcation, starting for instance from a stable equilibrium $(x^*, p^*) = (3.0,\ 0.6)$ to a new position, say $(x^o, p^o) = (4.0,\ 0.6)$, as shown in Fig. 6.8.

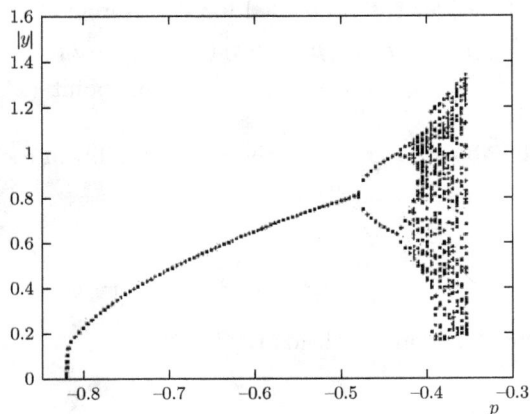

(a) before control

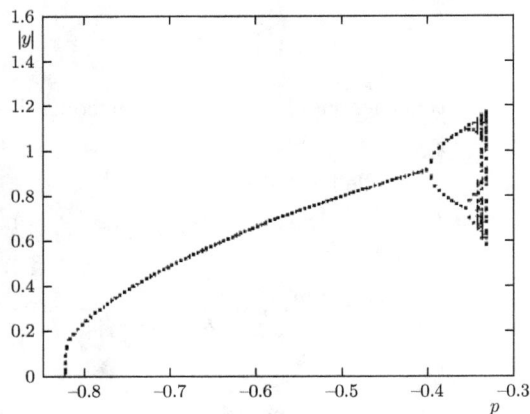

(b) after control

Fig. 6.8   Shifting the period-doubling bifurcation of the logistic map.

6.4.1.4   *Controlling Hopf Bifurcation*

Consider a controlled system:

$$\dot{x} = f(x, y; p),$$
$$\dot{y} = g(x, y; p) + u(x, y; p).$$

The objective is to design a simple controller, $u(x, y; p)$, which does not change the original equilibrium (x^*, y^*) but can move the original Hopf bifurcation point (x^*, y^*, p^*) to a new position $(x^o, y^o, p^o) \neq (x^*, y^*, p^*)$.

Clearly, the controller must satisfy $u(x^*, y^*; p) = 0$ for all p, in order not to change the position of (x^*, y^*).

The controlled system Jacobian at (x^o, y^o) is

$$J(p) = \begin{bmatrix} f_x & f_y \\ g_x + u_x & g_y + u_y \end{bmatrix}_{x=x^*, y=y^o=0}$$

where $f_x = \partial f / \partial x$ and $g_y = \partial g / \partial y$, etc., which has eigenvalues

$$\lambda_{1,2}^c(p) = \tfrac{1}{2}(f_x + g_y + u_y)$$
$$\pm \tfrac{1}{2} \sqrt{(f_x + g_y + u_y)^2 - 4\left[f_x(g_y + u_y) - f_y(g_x + u_x)\right]}$$

where $f_x := f_x\big|_{x=x^o, y=y^o}$, $g_y := g_y\big|_{x=x^o, y=y^o}$, etc.

To obtain a Hopf bifurcation at $(x^o, y^o; p^o)$, the above analysis leads to the following conditions:

(i) (x^o, y^o) is an equilibrium point of the controlled system:

$$f(x^o, y^o; p) = 0, \quad g(x^o, y^o; p) + u(x^o, y^o; p) = 0, \quad \forall\, p \in R;$$

(ii) the complex conjugate eigenvalues $\lambda_{1,2}^c(p)$ are purely imaginary at $(x^o, y^o; p^o)$:

$$(f_x + g_y + u_y)\Big|_{p=p^o} = 0,$$

$$f_x(g_y + u_y) - f_y(g_x + u_x)\Big|_{p=p^o} > 0,$$

$$(f_x + g_y + u_y)^2 - 4\left[f_x(g_y + u_y) - f_y(g_x + u_x)\right]\Big|_{p \neq p_0} < 0;$$

(iii) the crossing of the eigenlocus at the imaginary axis is transversal:

$$\frac{\partial \mathrm{Re}\{\lambda_1^c(p)\}}{\partial p}\Big|_{p=p^o} = \frac{\partial (f_x + g_y + u_y)}{\partial p}\Big|_{p=p^o} > 0.$$

Example 6.9. Consider a discrete-time controlled system:

$$x_{k+1} = f(x_k; p) + u_k(y_k; p),$$
$$y_{k+1} = x_{k+1} - x_k.$$

The controller is designed such that $u_k(0; p) = 0$, so it will not change the position of the original equilibrium x^*.

The Jacobian of the controlled system at $(x^o, y^o) = (x^*, 0)$ is

$$J(p) = \begin{bmatrix} f_x & u_y \\ f_x - 1 & 0 \end{bmatrix}_{x=x^*, y=y^o=0},$$

where $f_x = \partial f / \partial x_k$ and $u_y = \partial u_k / \partial y_k$, which has eigenvalues

$$\lambda_{1,2}(p) = \frac{1}{2} f_x \pm \frac{1}{2} \sqrt{f_x^2 - 4u_y(f_x - 1)}\,.$$

Conditions for the controller are

$$\lambda_1(p) = \bar{\lambda}_2(p)\,, \quad |\lambda_{1,2}(p^*)| = 1\,,$$

$$f_x^2 \le 4u_y(f_x - 1)\,, \quad \frac{\partial |\lambda_{1,2}(p^*)|}{\partial p} > 0\,.$$

As a specific example, the controlled logistic system

$$x_{k+1} = p\, x_k \left(1 - x_k\right) + u_k(y_k; p)\,,$$
$$y_{k+1} = x_{k+1} - x_k\,,$$

yielding $u_k = e^p$ and $y_k = e^p\,(x_k - x_{k-1})$. The control performance is visualized by a simulation shown in Fig. 6.9.

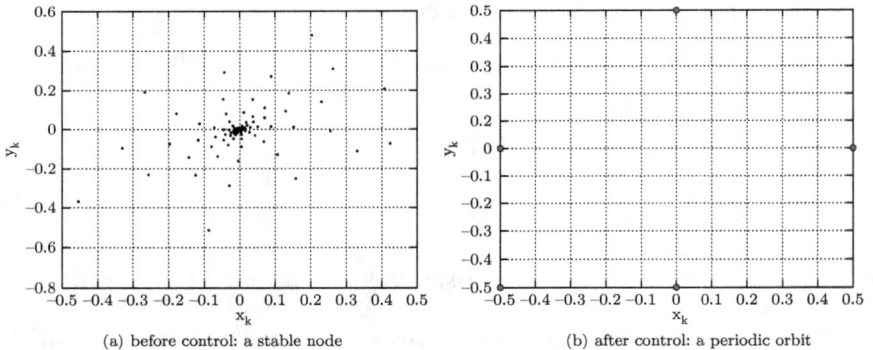

(a) before control: a stable node   (b) after control: a periodic orbit

Fig. 6.9   Controlling the logistic map from a stable node to a periodic orbit.

Example 6.10. Consider a special form of the Lorenz system:

$$\dot{x} = -p(x - y)\,,$$
$$\dot{y} = -xz - y\,,$$
$$\dot{z} = xy - z - r\,,$$

where p and r are parameters, which have three equilibria:

$$C_0 = (0, 0, 0)\,, \quad C_\pm = \pm\sqrt{r - 1}\,.$$

With $p = 4$:

• when $15.15 \leq r \leq 15.25$, the system has two locally stable equilibria $C_\pm$ and one chaotic attractor;

• when $r_H = 16$, the system has a Hopf bifurcation, as shown in Fig. 6.10, which shows the trajectories of the uncontrolled Lorenz system with $p = 4$, initial condition $(x_0, y_0, z_0) = (10.0, -10.0, 5.7)$: (a) chaos for $r = 15.15$; (b) stable C_- for $r = 13.20$; (c) stable C_+ for $r = 15.25$.

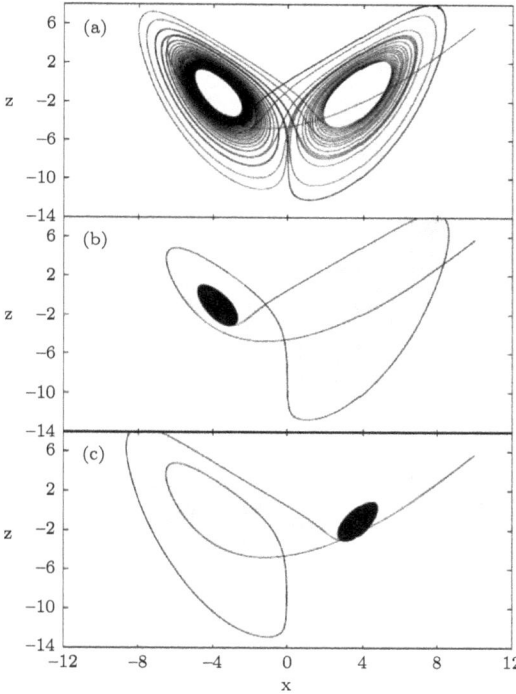

Fig. 6.10   Trajectories of the uncontrolled Lorenz system.

Now, design a feedback controller, named "washout filter", as follows:

$$\dot{x} = -p(x - y) \,,$$
$$\dot{y} = -xz - y \,,$$
$$\dot{z} = xy - z - r - u \,,$$
$$u = -k_c(y - cv) - k_n(y - cv)^3 \,,$$
$$\dot{v} = y - cv \,,$$

where $c = 0.5$, $k_c = 2.5$, $k_n = 0.09$, which yields $r_H = 36.004$ for C_+ and $r_H = 1.085$ for C_- (comparing to the original value of $r_H = 16$).

The trajectories of the Lorenz system under control with a washout filter are shown in Fig. 6.11, as follows:

(a) $k_n = 0.009$, converging to C_+ ;
(b) $k_n = 0.009$, diverging to ∞ ;
(c) $k_n = 10$, converging to a limit cycle;
(d) $k_n = 10$, converging to a limit cycle;
(e) $k_n = 10$, converging to C_- ;
(f) $k_n = 10$, converging to a limit cycle.

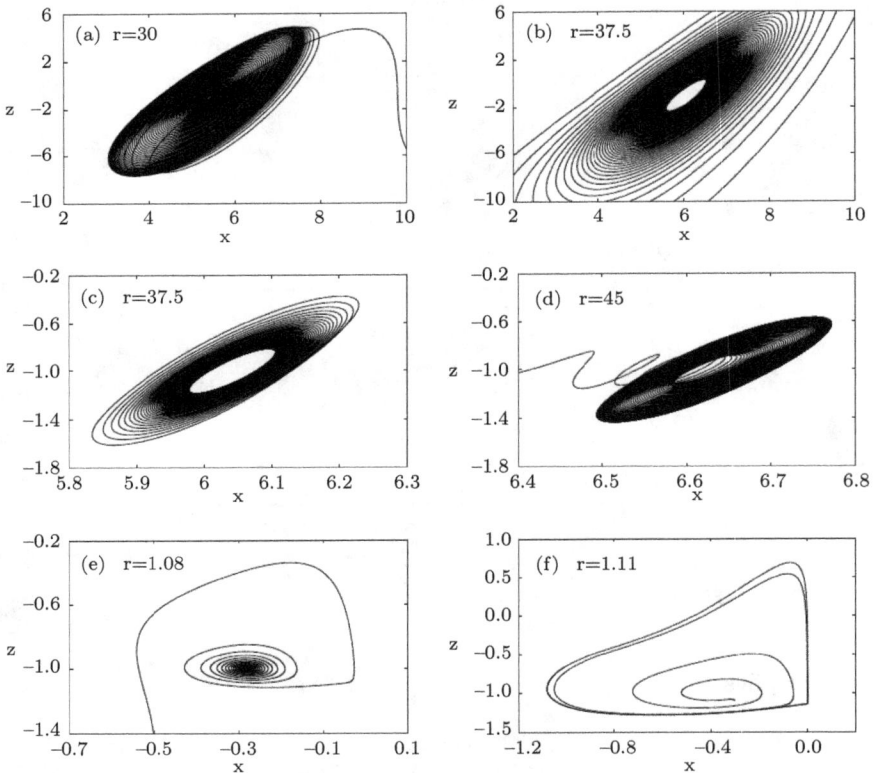

Fig. 6.11   Trajectories of the controlled Lorenz system.

6.4.1.5 *More about Bifurcation Control*

It is worth mentioning that bifurcation control is a rich subject to study, *albeit* fairly challenging, with, for instance, the following topics to be further investigated:

- delaying the onset of a bifurcation;
- stabilizing a bifurcation solution or branch;
- changing a bifurcation value;
- enlarging the parameter range of a bifurcation;
- modifying the shape or type of a bifurcation;
- introducing a new bifurcation point;
- creating a new type of bifurcation;
- reversing a period-doubling bifurcation;
- monitoring a degenerate Hopf bifurcation;
- controlling the multiplicity, amplitude, and/or frequency of limit cycles emerging from Hopf bifurcation;
- and, a certain combination of some of these.

6.4.2 *Controlling Chaos*

It was shown in Examples 6.1 and 6.2 that chaos can be controlled to periodic orbits, including equilibria as special cases. However, it should be noted that in these examples no specific chaos properties were utilized, where the chaotic systems were treated just like nonchaotic nonlinear systems. Therefore, using skillfully designed or even brute-force controllers it is often possible to accomplish stabilization or tracking control tasks.

In contrast, making a nonchaotic system to be chaotic by means of control is much harder. Since chaos can be useful in some engineering applications, such as liquids-mixing, information encryption, secure communication, weak signal detection, global optimality searching, etc., there is a need to generate chaos purposefully when it is desirable. This reverse task is referred to as *chaotification* (or, *anti-control of chaos*).

To do so, consider a general finite-dimensional discrete-time dynamical system, need not be originally chaotic or unstable, in the form of

$$x_{k+1} = f_k(x_k), \qquad x_0 \in R^n, \qquad (6.46)$$

where, at the moment, $f_k(\cdot)$ is only assumed to be continuously differentiable, at least locally in a region of interest. This means that the given system (6.46) can be linear or nonlinear, time-invariant or time-varying, and stable or unstable.

The objective is to design a controller in the form of a control input sequence, $\{u_k\}$, such that the output state vector sequence of the controlled system

$$x_{k+1} = f_k(x_k) + u_k \qquad (6.47)$$

behaves chaotically, in the sense of Devaney (Definition 5.6) or Li–Yorke (Definition 5.5)

As discussed in Sec. 6.1.1, a practical design of a useful controller should yield a simple structure, such that the goal of control (here, chaotification) can be achieved. The following discussions will follow this basic engineering principle, thereby designing some simple and implementable chaotifiers (anti-controllers).

6.4.2.1 Chaotification Problem Description

Consider a simple nonlinear state-feedback controller of the form

$$u_k = a_k\,\varphi(\sigma_k x_k)\,\mathbf{1}, \qquad (6.48)$$

where $\mathbf{1} = [1, \cdots, 1]^\top$ is an all-one vector, $\{a_k\}$ and $\{\sigma_k\}$ are two sequences of real constants to be determined, and $\varphi(\cdot)$ is a simple unimodal function such as a piecewise-linear sawtooth function, as shown in Fig. 6.12, or a sine function, both of which can be easily implemented by commercial circuits.

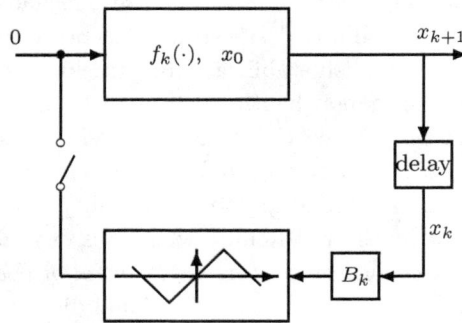

Fig. 6.12   The state-feedback anti-control configuration.

Using this controller, without tuning any of the system parameters, the controlled system

$$x_{k+1} = f_k(x_k) + a_k\,\varphi(\sigma_k x_k)\,\mathbf{1} \qquad (6.49)$$

is expected to become chaotic, regardless of the functional form of the given system $f_k(\cdot)$, which is only assumed to be Lipschitz so that its Jacobian can be uniformly bounded.

When the sawtooth function is used in (6.48), the controlled system is equivalent to the following:

$$x_{k+1} = f_k(x_k) + B_k x_k \quad (\text{mod } 1), \tag{6.50}$$

where $\{B_k\}$ is a sequence of constant matrices and (mod 1) is the component-wise modulo-one operation, which maps $[0,1]$ onto itself.

It is remarked that this equivalence is not obvious, but can be verified component-wise, namely in the 1-dimensional setting, as demonstrated by the simple example discussed at the end of this section below.

Note also that in operation (mod 1), the magnitude 1 can be replaced by any positive real number, such as 2π if the sine function is used in (6.48).

Thus, the controlled system (6.50) rather than (6.49) is discussed here for notational simplicity.

To describe the problem more precisely, some new notation is introduced. Let

$$J_j(z) = f_j'(z) + B_j \tag{6.51}$$

be the Jacobian of the controlled system (6.50), evaluated at z, $j = 0, 1, \ldots,$ and let

$$T_j = T_j(x_0, \ldots, x_j) := J_j(x_j) \cdots J_0(x_0).$$

Moreover, let μ_i^j be the ith eigenvalue of the jth product matrix $[T_j^\top T_j]$, $i = 1, \ldots, n$, $j = 0, 1, \ldots,$ and recall that the ith Lyapunov exponent of the orbit $\{x_k\}$ of the controlled system, starting from x_0, is given by (5.14), namely,

$$\lambda_i(x_0) = \lim_{k \to \infty} \frac{1}{2k} \ell n \, |\mu_i|, \quad i = 1, \ldots, n. \tag{6.52}$$

The chaotification approach comprises two steps:

First, design the constant matrices $\{B_k\}$ such that all the Lyapunov exponents of the controlled system orbit starting from x_0 are finite and strictly positive:

$$0 < c \le \lambda_i < \infty, \quad i = 1, \ldots, n, \tag{6.53}$$

where c is a pre-desired constant. To be practical in achieving this goal, namely, for implementation purpose, it is also required that the constant gain sequence $\{B_k\}$ is uniformly bounded for all $k = 0, 1, \ldots$. As a matter of fact, this can be guaranteed by using a very simple constant feedback gain $B_k \equiv \sigma I_n$ with some $\sigma > 0$ for all $k = 0, 1, \ldots$, as demonstrated below.

Second, take the (mod 1) operation, so as to force the diverging system orbits to return back into a bounded region, thereby generating chaos as desired.

It is remarked that the above-described procedure of chaotification is always possible under one (and only one) natural condition that all Jacobians $J_k(x_k)$ of the given system are uniformly bounded:

$$||f_k'(x_k)|| \leq M < \infty, \quad \text{for all } k = 0, 1, \ldots, \quad (6.54)$$

where any suitable matrix norm such as the spectral norm can be used. This restriction simply avoids that the system orbits diverge too quickly before the controller takes effective action. This condition is not too restrictive, since many systems such as all linear time-invariant systems satisfy it.

6.4.2.2   *A General Chaotification Algorithm*

Start with the initial controlled system $x_1 = f_0(x_0) + B_0 x_0$, where x_0 is given. Calculate its Jacobian $J_0(x_0) = f_0'(x_0) + B_0$ and set $T_0 = J_0(x_0)$. Design a positive feedback control gain, $B_0 = \sigma_0 I$, by choosing a constant $\sigma_0 > 0$ such that the matrix $[T_0 T_0^\top]$ is finite and diagonally dominant.

For $k = 0, 1, 2, \ldots$, start with the controlled system $x_{k+1} = f_k(x_k) + B_k x_k$, where both B_k and x_k are obtained from the previous step. Perform the following steps iteratively:

Step 1. Compute the Jacobian $J_k(x_k) = f_k'(x_k) + \sigma_k I$ and then set $T_k = J_k T_{k-1}$.

Step 2. Design a positive feedback controller by choosing a constant $\sigma_k > 0$ such that the matrix $[T_k T_k^\top - e^{2kc} I]$ is finite and diagonally dominant, where the constant $c > 0$ is the one specified in (6.53).

Step 3. Apply the (mod 1) operation to the controlled system, as shown in (6.50), or apply a unimodal function (e.g. the piecewise-linear sawtooth function shown in Fig. 6.12) to the controller, as formulated in (6.49).

To this end, it can be rigorously proved [Chen (2003)] that the controlled system so designed is guaranteed to be chaotic in the sense of Devaney and also of Li–Yorke, as further illustrated below.

A few remarks are in order.

First, a simple choice for $\{\sigma_k\}$ is to use the constant $\sigma = M + e^c$, where c is given in (6.53) and M is given in (6.54), for all $k = 0, 1, \ldots$.

Second, in Step 3 above, if the sawtooth function is used, its magnitude can be a constant, $a_k = a$ for all $k = 0, 1, \ldots$, and this constant can be arbitrarily small, which means that "small control" is possible; but in this case, σ should be large, i.e. the frequency should be high.

Third, some other unimodal functions such as the sine function can also be used at Step 3 for the controller.

It is particularly important to emphasize that this chaotification scheme is fairly general because the given system (6.46) can be quite arbitrary: it can be linear or nonlinear, time-invariant or time-varying, stable or unstable, of any finite dimension, provided that the system Jacobians are uniformly bounded in the sense of (6.54).

It is easily noticed that the above chaotification scheme utilizes a full-state feedback, which is often not desirable in engineering applications because full-state information is not always available, e.g. using sensors in practice. For engineering control systems, it is more preferable to use output feedback instead of state feedback, which means that only part of the state vector is measured and used for feedback control. Therefore, it is natural to modify the above algorithm and use output feedback instead of full-state feedback to achieve chaotification, which is further discussed next.

6.4.2.3 *A Modified Chaotification Algorithm*

Recall the chaotifier (6.48), namely,

$$u_k = a_k \, \varphi(\sigma_k x_k) \, \mathbf{1}, \tag{6.55}$$

where $\mathbf{1} = [1, \cdots, 1]^\top$ is an all-one vector, $\{a_k\}$ and $\{\sigma_k\}$ are two sequences of real constants to be determined, and $\varphi(\cdot)$ is a simple unimodal function as the piecewise-linear sawtooth function or the sine function.

Now, this full-state feedback chaotifier is modified to be

$$u_k = b\varphi(x_k) = -b \sin(h x_k), \tag{6.56}$$

where b is an $n \times 1$ constant vector and h is a $1 \times n$ constant vector, both are to be determined.

The chaotifier (6.48) can also be modified as

$$u_k = a\varphi(x_k) = a \left[\sin x_k^1 \ \sin x_k^2 \ \cdots \ \sin x_k^n \right]^\top, \tag{6.57}$$

where a is a constant to be determined, and $x_k = \left[x_k^1 \ x_k^2 \ \cdots \ x_k^n \right]^\top$ is the state vector.

The above anti-controller can also be slightly modified to be

$$u_k = a\phi(cx_k) = a \left[\sin cx_k^1 \ \sin cx_k^2 \ \cdots \ \sin cx_k^n \right]^\top, \tag{6.58}$$

where a and c are constants to be determined and $x_k = \begin{bmatrix} x_k^1 & x_k^2 & \cdots & x_k^n \end{bmatrix}^{\top}$ is the state vector as above.

Note that the corresponding controlled system (6.47) using either of these chaotifiers is chaotic in the sense of Devaney for some specifically given systems, as demonstrated by the following example.

6.4.2.4   *An Illustrative Example*

To illustrate how the above-discussed state-feedback (or output-feedback) controller can chaotify a given nonchaotic system, and how to verify that the controlled system is chaotic, a simple 1-dimensional example is discussed.

Consider the 1-dimensional linear state-feedback control system

$$x_{k+1} = a\, x_k + u_k \qquad (\text{mod } 1) \tag{6.59}$$

where the controller

$$u_k = \left(N + e^c\right) x_k\,, \tag{6.60}$$

with $0 < c \le |a| \le N < \infty$.

Putting together, this controlled system is

$$x_{k+1} = a\, x_k + \left(N + e^c\right) x_k = \left(a + N + e^c\right) x_k \quad (\text{mod } 1) \tag{6.61}$$

where $\left(a + N + e^c\right) > 1$.

Define a map $\phi : S^1 \to S^1$, where S^1 is the unit circle on the 2-dimensional plane, by

$$\phi(x) = \left(a + N + e^c\right) \angle x\,, \quad x \in S^1\,, \tag{6.62}$$

in which $\angle x$ is the angle of the point $x \in S^1$ (thus, $0 \le \angle x < 2\pi$ and ϕ is 2π-periodic). Note that this map is a one-variable map since the radius of the circle is fixed and only the angle is variable.

It can be verified that the two systems, (6.61) and (6.62), are equivalent. Indeed, by multiplying 2π on both sides, it can be seen that (6.61) is equivalent to

$$x_{k+1} = \left(a + N + e^c\right) x_k \quad (\text{mod } 2\pi)\,, \tag{6.63}$$

or

$$\angle x_{k+1} = \left(a + N + e^c\right) \angle x_k\,, \qquad x_k \in S^1\,,$$

which is equivalent to (6.62) if the initial state x_0 can be arbitrary.

Finally, it can be shown that the map (6.63) is equivalent to the chaotic logistic map $x_{k+1} = 4x_k(1-x_k)$ by defining $x_k = \frac{1}{2}\left[1 - \cos(2\pi\xi y_k)\right]$ in the logistic map, where $\xi := (a + N + e^c) > 1$. With this change of variables, the logistic map becomes

$$\frac{1}{2}\left[1 - \cos(2\pi\xi y_{k+1})\right] = 4 \cdot \frac{1}{2}\left[1 - \cos(2\pi\xi y_k)\right] \cdot \left\{1 - \frac{1}{2}\left[1 - \cos(2\pi\xi y_k)\right]\right\}$$
$$= \left[1 - \cos(2\pi\xi y_k)\right] \cdot \left[1 + \cos(2\pi\xi y_k)\right],$$

which has the solution precisely given by formula (6.63).

Exercises

6.1 Consider the pendulum model

$$\begin{cases} \dot{x} = y \\ \dot{y} = -\frac{\kappa}{m} y - \frac{g}{\ell} \sin(x) + u, \end{cases}$$

where u is a variable parameter used as an adjustable constant control input. Convert this system into the normal form.

6.2 * Consider the following controlled system:

$$\dot{\mathbf{x}} = A\mathbf{x} + \mathbf{f}(\mathbf{x}, t) + \mathbf{u}(\mathbf{x}, t),$$

where A is a stable constant matrix and

$$\lim_{||\mathbf{x}|| \to \infty} \frac{||\mathbf{f}(\mathbf{x}, t)||}{||\mathbf{x}||} = 0 \qquad \text{for all } t \geq t_0.$$

Suppose that the controller $\mathbf{u}$ is designed such that

$$||\mathbf{u}(\mathbf{x}, t)|| \leq \gamma(t) \qquad \text{for all } t \geq t_0 \text{ and } ||\mathbf{x}|| \leq \kappa,$$

for some constant $\kappa < \infty$, where $\gamma(t)$ is a non-negative continuous function satisfying for any constant $a > 0$, that

$$\int_t^{t+a} \gamma(\tau) d\tau \to 0 \qquad (t \to \infty).$$

Show that, for the controlled system, $\mathbf{x}(t) \to 0$ as $t \to \infty$.

6.3 Consider an uncertain nonlinear control system,

$$\begin{cases} \dot{x} = y \\ \dot{y} = f(\mathbf{x}) + u, \end{cases}$$

where $\mathbf{x} = [x \ y]^\top$ and f is unknown but satisfies $|f(\mathbf{x})| \leq c||\mathbf{x}||$ for some constant $c > 0$, and the controller

$$u = -x - y + v,$$

with v to be determined. Verify that if v is designed as

$$v = \begin{cases} -c||\mathbf{x}|| \frac{w}{|w|} & \text{if } c||\mathbf{x}|| \cdot |w| \geq \varepsilon \\ -c^2||\mathbf{x}||^2 \frac{w}{\varepsilon} & \text{if } c||\mathbf{x}|| \cdot |w| < \varepsilon, \end{cases}$$

where $w = 2\mathbf{x}^\top P \begin{bmatrix} 0 \\ 1 \end{bmatrix}$ with P being a positive definite and symmetrical matrix, then the uncertain system can be stabilized to zero, namely, $x \to 0$ and $y \to 0$ as $t \to \infty$.

[Hint: Consider the Lyapunov function $V = \mathbf{x}^\top P\mathbf{x}$.]

6.4 Consider the motion equation of a controlled single-link robot arm:

$$D(q)\ddot{q} + C(q,\dot{q})\dot{q} + \phi(q) = u,$$

where $q = q(t)$ is the joint angular variable, $C > 0$, $D > 0$ and ϕ are nonlinear matrix- or vector-valued functions, and $u = u(t)$ is the controller input to be designed. The objective is to control the system output $q(t)$ to track a target trajectory $q^d = q^d(t)$ as $t \to \infty$. Suppose that the controller $u(t)$ is designed in three steps:

i. Assume a control input $u(t)$ of the form

$$u = D_0(q)v + C_0(q,\dot{q})\dot{q} + \phi(q),$$

where $c = c(t)$ is a new controller to be further designed. Verify that, by designing $\mathbf{x} = [x_1 \ x_2]^\top = [q \ \dot{q}]^\top$, the controlled system can be reformulated as

$$\dot{x} = A\mathbf{x} + B(v + \xi),$$

and find the explicit expressions of the two constant matrices A and B, as well as the uncertain term $\xi = \xi(\mathbf{x})$, which contains all the other terms including perhaps also $\mathbf{x}$.

ii. Introduce the error signal $\mathbf{e} = [e_1 \ e_2]^\top$ with

$$\begin{cases} e_1 = x_1 - x_1^d = q - q^d \\ e_2 = x_2 - x_2^d = \dot{q} - \dot{q}^d. \end{cases}$$

Show that the above error-signal model can be reformulated as

$$\dot{\mathbf{e}} = A_e\mathbf{e} + B_e(v + \xi_e),$$

and find the explicit expressions of the two constant matrices A_e and B_e, as well as the uncertain term $\xi_e = \xi_e(\mathbf{x})$, which contains all uncertain terms including perhaps q^d, $\dot{q}^d$ and $\ddot{q}^d$. Is this error-signal model stable?

iii. Suppose that the model uncertainties ΔC and $\Delta\phi$ are both bounded, although unknown, in the sense that $|\xi_e| \leq 1$. Design the controller by determining a controller

$$v = \mathbf{k}^\top\mathbf{e} + v_0 = [k_1 \ k_2]\begin{bmatrix} e_1 \\ e_2 \end{bmatrix} + v_0$$

for some constant feedback control gains k_1 and k_2, as well as a constant control input v_0, such that the above controlled error-signal model is stable with $\mathbf{e}(t) \to 0$ as $t \to \infty$. If this is impossible, explain what the basic limitation of a linear controller is for an uncertain nonlinear system, and suggest a simple working nonlinear (e.g. adaptive) controller for this robot arm system.

6.5 Verify that the reaching law (6.41) satisfies the reaching condition (6.40).

6.6 In Example 6.8, it was shown that a linear controller cannot produce a pitchfork bifurcation. Verify that an additional cubic term to the linear controller can do the job.

6.7 Consider the simplified pendulum model

$$\ddot{\theta} + a\,\dot{\theta} + b\,\sin(\theta) = u\,,$$

where $0 < a < b$ are given constant parameters. Verify, by both mathematical analysis and physical interpretation, that the uncontrolled pendulum with $u = 0$ does not have a Hopf bifurcation. Then, design a linear controller of the form

$$u = \mu(\theta + \dot{\theta})\,,$$

with $\mu = \mu(a,b)$ being a suitably chosen variable control gain, such that the controlled pendulum has a Hopf bifurcation.

6.8 * Verify that the controlled system described by the map (6.63) satisfies Definition 5.6 for chaos:

  i. Verify that, after k iterations of the map, one has $\phi^k(x) = \xi^k \angle x$.

  ii. Since ϕ is 2π-periodic, let $\xi^k \angle x = \angle x + 2\ell\pi$, where ℓ is an integer satisfying $0 \le \ell \le \lfloor \xi^k \rfloor - 1$. Verify that the periodic solutions of the map ϕ are almost uniformly distributed on the unit circle:

$$\angle x_{k,\ell} = \frac{2\ell\pi}{\xi^k - 1}, \quad \ell = 0, 1, \ldots, \lfloor \xi^k \rfloor - 1\,,$$

which means that for two arbitrarily given nonempty open subsets in S^1, there are two large enough integers k_0 and ℓ_0 such that the periodic solution $\{x_{k_0,\ell_0}\}$ of the map has at least one point in each of these two subsets.

Bibliography

Anderson, B. D. O. and Vongpanitterd, S. [1972] *Network Analysis and Synthesis: A Modern System Approach* (Prentice-Hall).

Arrowsmith, D. K. and Place, C. M. [1990] *An Introduction to Dynamical Systems* (Campridge Univ. Press).

Chen, G. [1999] "Stability of nonlinear systems," in *Encycropidia of E E* (W. K. Chen, ed.) (Wiley) 2004 Edition: pp. 4881–4896.

Chen, G. [2003] "Chaotification via feedback: The discrete case," in *Chaos Control* (G. Chen and X. Yu, eds.) (Springer), pp. 159–177.

Chen, G. [2020] "Generalized Lorenz systems family," https://arxiv.org/abs/2006.04066

Coddington, E. A. and Levinson, N. [1955] *Theory of Ordinary Differential Equations* (McGraw-Hill).

Desoer, C. A. and Vidyasagar, M. [1975] *Feedback Systems: Input-Output Properties* (Academic Press).

Devaney, R. L. [1987] *An Introduction to Chaotic Dynamical Systems* (CRC Press).

Glendinning, P. [1994] *Stability, Instability and Chaos: An Introduction to the Theory of Nonlinear Differential Equations* (Cambridge Univ. Press).

Hahn, W. [1967] *Stability of Motion* (Springer-Verlag).

Hoppensteadt, F. C. [2000] *Analysis and Simulation of Chaotic Systems* (Springer).

Huygens, C. [1665] "A letter to the Royal Society," in *Oeuvres Completes de Christian Huygens*, M. Nijhoff (ed.), Societe Hollandaise des Sciences, The Hague, The Netherlands, 1893, vol. 5, p. 246 (in French).

Khalil, H. K. [1996] *Nonlinear Systems* (Prentice-Hall).

Li, T. Y. and Jorke, J. A. [1975] "Period three implies chaos," Amer. Math. Month., 82, 481–485.

Marotto, F. R. [1978] "Snap-back repellers imply chaos in R^n," J. Math. Anal. Appl., 63, 199–223.

Massera, J. L. [1956] "Contributions to stability theory," Annals of Mathematics, 64(1): 182–206.

Moiola, J. L. and Chen, G. [1996] *Hopf Bifurcation Analysis: A Frequency Domain Approach* (World Scientific).

Narendra, K. S. and Taylor, J. H. [1973] *Frequency Domain Criteria for Absolute Stability* (Academic Press).

Popov, V. M. [1973] *Hyperstability of Automatic Control Systems* (Springer-Verlag).

Robinson, C. [1995] *Dynamical Systems: Stability, Symbolic Dynamics, and Chaos* (CRC Press).

Sastry, S. [1999] *Nonlinear Systems: Analysis, Stability and Control* (Springer).

Touhey, P. [1997] "Yet another definition of chaos," Amer. Math. Month., 104, 411–414.

Verhulst, F. [1996] *Nonlinear Differential Equations and Dynamical Systems* (Springer).

Ważewski, T. [1950] "Systémes des équations et des inégalités différentielles ordinaires aux deuxiéme members monotones et leurs applications," Ann. Soc. Polon. Math., 23, 112–166 (in French).

Wiggins, S. [1990] *Introduction to Applied Nonlinear Dynamical Systems and Chaos* (Springer).

Index